Sensory Evaluation
A practical handbook

To George, Elizabeth, George and William
To Mike, Holly and Socks
To Campbell, Emma and Lara

Sensory Evaluation
A practical handbook

Dr Sarah E. Kemp
Former Head of Global Sensory and Consumer Guidance
Cadbury Schweppes PLC
UK

Dr Tracey Hollowood
Associate Director: Sensory and Consumer Research
Sensory Dimensions Ltd
Bulwell
Nottinghamshire
UK

Dr Joanne Hort
Associate Professor in Sensory Science
University of Nottingham
Sutton Bonington
Leicestershire
UK

A John Wiley & Sons, Ltd., Publication

This edition first published 2009
© 2009 S.E. Kemp, T. Hollowood and J. Hort

Blackwell Publishing was acquired by John Wiley & Sons in February 2007. Blackwell's publishing programme has been merged with Wiley's global Scientific, Technical, and Medical business to form Wiley-Blackwell.

Registered office
John Wiley & Sons Ltd, The Atrium, Southern Gate, Chichester, West Sussex, PO19 8SQ, United Kingdom

Editorial offices
9600 Garsington Road, Oxford, OX4 2DQ, United Kingdom
2121 State Avenue, Ames, Iowa 50014-8300, USA

For details of our global editorial offices, for customer services and for information about how to apply for permission to reuse the copyright material in this book please see our website at www.wiley.com/wiley-blackwell.

The right of the author to be identified as the author of this work has been asserted in accordance with the Copyright, Designs and Patents Act 1988.

All rights reserved. No part of this publication may be reproduced, stored in a retrieval system, or transmitted, in any form or by any means, electronic, mechanical, photocopying, recording or otherwise, except as permitted by the UK Copyright, Designs and Patents Act 1988, without the prior permission of the publisher.

Wiley also publishes its books in a variety of electronic formats. Some content that appears in print may not be available in electronic books.

Designations used by companies to distinguish their products are often claimed as trademarks. All brand names and product names used in this book are trade names, service marks, trademarks or registered trademarks of their respective owners. The publisher is not associated with any product or vendor mentioned in this book. This publication is designed to provide accurate and authoritative information in regard to the subject matter covered. It is sold on the understanding that the publisher is not engaged in rendering professional services. If professional advice or other expert assistance is required, the services of a competent professional should be sought.

A catalogue record for this book is available from the British Library and Library of Congress.

ISBN: 978-1-4051-6210-4

Set in 9.5/12 pt Minion by Macmillan Publishing Solutions
1 2009

Contents

Preface	vii
Author biographies	ix
Acknowledgements	xi

1 Introduction — 1
1.1 What is sensory evaluation? — 1
1.2 What is the role of sensory evaluation? — 2
1.3 What drives successful sensory testing? — 3

2 Sensory perception — 4
2.1 The human senses — 4
2.2 Factors affecting sensory measurements — 6

3 Planning your sensory project — 11
3.1 Setting objectives — 11
3.2 Product type — 11
3.3 Budget — 12
3.4 Timings — 12
3.5 Selecting the test method — 12
3.6 Setting action standards — 13
3.7 Experimental design — 14
3.8 Data analysis — 19

4 Requirements for sensory testing — 30
4.1 Professional conduct in sensory testing: health, safety, ethical and legal considerations — 30
4.2 Good working and laboratory practices — 37
4.3 Resources needed for sensory testing — 41
4.4 Samples — 49
4.5 Assessors — 54
4.6 Data capture — 63

vi **Contents**

5 Sensory test methods — 66
5.1 Selecting the test — 66
5.2 Discrimination tests — 66
5.3 Descriptive analysis tests — 96
5.4 Affective/consumer tests — 118
5.5 Linking consumer, sensory and product data — 136

6 Completing the project — 138
6.1 Reporting — 138
6.2 Documentation and data storage — 140
6.3 Dos and don'ts — 141

7 Appendices — 142
Appendix 1: Examples of Latin Square and Williams Latin Square designs for selected number of samples — 142
Appendix 2: IFST PFSG professional code of conduct for sensory professionals — 143
Appendix 3: Critical values table for triangle test — 147
Appendix 4: Critical values table for duo-trio test and paired comparison test for difference (one tailed) — 149
Appendix 5: ANOVA explained — 151
Appendix 6: Critical values table for chi-squared — 156
Appendix 7: Critical values table for paired comparison and paired difference test (two tailed) — 157
Appendix 8: Critical values table for Friedman test — 159
Appendix 9: Types of scales — 160
Appendix 10: Case study: modified quantitative descriptive analysis of chocolate texture — 163
Appendix 11: R index explained — 174

8 Glossary — 178

9 References — 185

Index — 189

Preface

This book is an affordable sensory science textbook focused on the practical aspects of sensory testing on a broad range of products. It is presented in a simple 'how to' style for use by industry and academia as a step-by-step guide on how to carry out a range of sensory tests. It is intended as a companion volume to a larger, more detailed sensory science textbook covering theoretical aspects, advanced techniques and applications of sensory evaluation. The inspiration for this book is the excellent *Laboratory Methods for Sensory Evaluation of Food* by Elizabeth Larmont first published in 1967 and revised in 1977 and 1991 (Poste et al., 1991). It is now out of print; but at the time of publication, it was popular for its practical, easy-to-read style, coupled with good use of examples and illustrations. The authors have fond memories of using the book during their formative years in sensory science.

Between them, the authors have over 50 years of industrial and academic experience in sensory science and have published widely in the field. All three authors are founder committee members of The Institute of Food Science and Technology's Professional Food Sensory Group (IFST PFSG).

There are many good sensory textbooks on the market. The generalist sensory science texts are very comprehensive, but are often written in a research style, or with large sections of unbroken text which renders them unsuitable for use as a simple training/teaching aid or as a quick practical guide. They are also expensive/unaffordable in developing countries and difficult to understand for readers who have English as a second language. In addition, more and more specialised sensory texts are now available which tend to focus on theory and application in a narrow field, rather than general practice. There is a tendency for sensory textbooks to focus on food and beverage applications, often to the exclusion of other product categories.

The objectives of this book are as follows:
- To provide a practical guide and laboratory manual on how to carry out sensory evaluation techniques.

- To reach sensory practitioners, as well as sensory scientists, by using a simple, easy-to-read, easy-to-use format.
- To cover a broad range of product applications, including food, beverages, personal care and household products.
- To be inexpensive and available to a wide audience who would not usually be able to afford to purchase standard sensory textbooks, including students, technicians and practitioners in developing countries.
- To cover the IFST PFSG accreditation scheme at foundation and intermediate levels.

The very simple, practical, easy-to-use style of this book, coupled with its affordability, makes it suitable as a training manual, reference text, teaching aid and course book. Key audiences include sensory practitioners, junior sensory staff, sensory students and sensory trainers. It is applicable across a broad range of industries and to those with limited budgets.

The style of the book is easy-to-follow 'instructions' with simple explanations of how and why to do testing in a particular way, rather than detailed theory and underlying science of techniques. It is laid out in logical sequence. Examples and illustrations are used throughout. Practical tips and hints in the form of dos and don'ts are included in each section.

The book begins with an introductory chapter that gives an overview of sensory evaluation and a second chapter on sensory perception. The third chapter outlines how to plan a sensory project. The fourth chapter focuses on requirements for sensory testing. Important elements of this chapter are professional conduct and good laboratory practice. These often receive scant coverage, but are becoming increasingly important as novel ingredients and processes continue to be developed (e.g. ingredients from genetically modified origin), and as products are increasingly tested in markets with regulations that are different from those in the markets for which they were designed. No matter how informal the sensory assessment is, it is essential that safe and ethical practices are used. The fifth chapter covers sensory test methods. Methods for statistical analysis are given throughout this chapter, rather than as a stand-alone section, to make the translation to practice easier. Case studies are used to illustrate methods. The sixth chapter covers elements necessary to complete a sensory project. Also included are appendices, glossary, references and index.

The authors hope that you enjoy using this book and that it helps bring success in your sensory endeavours.

Sarah E. Kemp
Tracey Hollowood
Joanne Hort

Author biographies

Sarah Elizabeth Kemp, *BSc (Hons), PhD, CSci, FIFST*, is a sensory and consumer science professional with more than 20 years of experience in academia and industry. Dr Kemp gained a BSc in Food Technology in 1986 and a PhD in Taste Chemistry in 1989 from the Food Science and Technology Department at Reading University, UK. In 1990, she did a postdoctoral research fellowship on sensory analysis at the Monell Chemical Senses Center in Philadelphia, USA. Dr Kemp has held numerous positions in the industry, including Manager of Sensory Psychology in the Fragrance Division of Givaudan-Roure in New Jersey, USA, Director of European Consumer and Marketing Research in the Fragrance Division at Givaudan-Roure, France, Product Area Leader and Sensory Science Leader in Foods Consumer Science at Unilever Research Colworth, UK, Former Head of Global Sensory and Consumer Guidance at Cadbury Schweppes, UK, and Director of Sensory and Consumer Services at Reading Scientific Services Ltd, UK. Dr Kemp has also set up and run her own consultancy service, Kemps Research Solutions Ltd. She has written numerous scientific articles in the field of sensory evaluation, has provided sensory training courses, including lecturing on the European Masters Course in Food Science, and has worked with bodies developing standards in sensory evaluation, such as the American Society for Testing and Materials. She is a founder member of the Professional Food Sensory Group of the Institute of Food Science and Technology, and a member of several other professional bodies, including the Sensory Evaluation Division of the Institute of Food Technologists, the Consumer and Sensory Research Technical Interest Group of the Society of Chemical Industry and the Association for Chemoreception Sciences.

Tracey Hollowood, *BSc (Hons), PhD, MIFST*, is currently Associate Director of Sensory and Consumer Research for Sensory Dimensions, UK. She has over 15 years of experience in academia and industry; she worked at Nottingham University for 10 years during which time she achieved her doctorate investigating perceptual taste–texture–aroma interactions. She established the United Kingdom's first Post Graduate Certificate in Sensory Science, and designed and managed the university's prestigious Sensory Science Centre. Her research has focused on psychophysical studies, interactions in sensory modalities and fundamental method development. She has over 20 peer-reviewed publications; has given numerous oral presentations and workshops; and has participated in the organisation of seven international symposia, including International Symposium of Taste 2000 and Pangborn Sensory Science Symposium 2005. She is the current Chair of the Institute of Food Science and Technology (IFST), Midland Branch and the Professional Food Sensory Group (PFSG).

Joanne Hort, *BEd (Hons), PhD, MIFST*, is Associate Professor in Sensory Science in the Division of Food Sciences at the University of Nottingham. Initially, she studied Food Technology and began her career in teaching. However, she returned to the university to receive her doctorate concerning the modelling of the sensory attributes of cheese from analytical and instrumental measures in 1998. As a lecturer at Sheffield Hallam University, she carried out sensory consultancy for local industry, developed a sensory program at undergraduate level and oversaw the installation of new sensory facilities before being appointed as Lecturer in Sensory Science at the University of Nottingham in 2002. She has since established the University of Nottingham Sensory Science Centre, which is renowned for both its sensory training and research into flavour perception. She delivers sensory courses at both undergraduate and postgraduate levels and is the Course Director for the Postgraduate Certificate in Sensory Science. Her research interests focus on the multimodal aspects of flavour perception and she has published several articles in this area, together with oral presentations and posters at international symposia. She is a founder member of the Professional Food Sensory Group of the Institute of Food Science and Technology and was on the organising committee of the 6th International Pangborn Symposium in the United Kingdom in 2005.

Acknowledgements

The authors would like to thank Simon Hails, of Cadbury plc and formally RSSL Ltd, for providing information on the ethical and legal considerations in sensory testing, Emma Louise Hewson for providing the diagram of a sensory testing facility and Rebecca Clark for assistance in compiling the references.

1 Introduction

It is estimated that 75% of new products fail within their first year on the supermarket shelf (Buisson 1995) and that, as a consequence, considerable resource invested in product development is squandered (Deschamps and Nayak 1996). Sensory attributes, whether the flavour of coffee, the smell of an air freshener, the texture of fabric or even the sound of a car door closing, are key determinants of product delivery including quality, functional and emotional benefits. Thus, a considerable proportion of product failure can be attributed to a mismatch between sensory properties and consumer needs or expectations. When integrated within the product development process, sensory and consumer testing allows cost-effective delivery of acceptable products to consumers and thus reduces the risk of failure (Lawless and Heymann 1998).

1.1 What is sensory evaluation?

Sensory evaluation is often described using the definition of Institute of Food Technology – a scientific method used to evoke, measure, analyse and interpret those responses to products as perceived through the senses of sight, smell, touch, taste and hearing (Anonymous 1975).

Since its emergence in the 1940s, however, sensory evaluation has developed as an exciting, dynamic, constantly evolving discipline that is now recognised as a scientific field in its own right.

The sensory professional is routinely confronted with problems which call upon an extensive skill set drawn from a range of disciplines, e.g. biological sciences, psychology, experimental design and statistics and will often be required to work with other specialists from these areas. Additional challenges are presented by working with a human 'measuring instrument' that is highly variable.

Sensory evaluation can be divided into two categories of testing: objective and subjective. In objective testing, the sensory attributes of a product are evaluated by a selected or trained panel. In subjective testing, the reactions of consumers to the sensory properties of products are measured. The power of sensory evaluation is realised when these two elements are combined to reveal insights into the way in which sensory properties drive consumer acceptance and emotional benefits. Linking sensory properties to physical, chemical, formulation and/or process variables then enables the product to be designed to deliver optimum or appropriate consumer benefits.

1.2 What is the role of sensory evaluation?

The role of sensory evaluation has changed considerably over the years. Initially, it was a service provider supplying data, but now its role is, in partnership with R&D and marketing, to provide insights to help guide development and commercial strategy.

From product conception to post-launch monitoring, sensory professionals can be called upon to inform decision-making during the stages of a product's life cycle. Sensory and consumer testing can also provide insights into human behaviour and perception at a more fundamental level.

In the early stages of product development, consumer and sensory testing can help identify the important sensory attributes driving acceptability across a product category. It can identify sensory-based target consumer segments, analyse competitor products and evaluate new concepts.

Combining data from sensory and instrumental testing may provide insights into the chemical and physical properties, driving sensory attributes. Where significant correlations exist with sensory data, it may be possible to dispense with the use of a sensory panel, in favour of a more cost-effective instrumental test, e.g. in quality testing.

Sensory testing can determine the impact of scaling up kitchen and/or pilot samples to large-scale production and is invaluable in determining whether raw ingredient changes or modifications to the production process, e.g. for cost reduction or change of supplier, will impact on sensory quality and/or product acceptability.

In terms of quality assurance, it can be used as part of a QA programme on raw materials. In addition, sensory testing can set consumer acceptability limits for sensory specifications used during quality testing. For those products susceptible to taints, sensory testing can ensure substandard products are not released onto the market. For many products, the sensory properties deteriorate ahead of microbial quality and so, in tandem with microbial tests, sensory testing can be used to determine shelf life and product variability through the supply chain.

From a marketing perspective, sensory and consumer testing can inform understanding concerning product preferences and acceptability. It can provide the data to support marketing claims such as 'best ever', 'new creamier', and 'most preferred'. It can also ensure that sensory properties work in synergy with brand communication and advertising.

Sensory and consumer testing is widely employed in the research arena. It is used at a more fundamental level to investigate new technologies to aid product development and to understand consumer behaviour. Furthermore, multidisciplinary investigations linking sensory testing with, for example instrumental analyses, brain-imaging techniques, psychophysical tests and genomics provide a wider understanding of the mechanisms involved in sensory perception and the variations that exist within the population.

1.3 What drives successful sensory testing?

Successful sensory testing is driven by setting clear objectives, developing robust experimental strategy and design, applying appropriate statistical techniques, adhering to good ethical practice and successfully delivering actionable insights that are used to inform decision-making. Appropriate training is crucial to ensure that the sensory professional has the necessary technical capability and interpersonal skills.

The aim of this book is to provide new and current sensory professionals with a firm foundation in the above principles in a practical, easy to follow format.

2 Sensory perception

2.1 The human senses

Sensory properties are perceived when our sensory organs interact with stimuli in the world around us. Consequently, it is important for sensory professionals to have some understanding of the biological mechanisms involved in perception. A basic outline of each sensory system is given in the following sections. *For more detailed information on the human senses, see Goldstein (2006).*

2.1.1 Vision

The appearance of any object is determined by the sense of vision. Light waves reflected by an object enter the eye and fall on the retina. The retina contains receptor cells, known as rods and cones, which convert this light energy into neural impulses that travel via the optic nerve to the brain. Cones are responsive to different wavelengths of light relating to 'colour'. Rods respond positively to white light and relay information concerning the lightness of the colour. The brain interprets these signals and we perceive the appearance (colour, shape, size, translucency, surface texture, etc.) of the object.

2.1.2 Gustation

The sense of taste involves the perception of non-volatile substances which, when dissolved in water, oil or saliva, are detected by taste receptors in the taste buds located on the surface of the tongue and other areas of the mouth or throat. The resulting sensations can be divided into five different taste qualities – salty, sweet, sour, bitter and umami. Examples of compounds that elicit particular tastes are given as follows:
- Salty substances: sodium chloride, potassium chloride
- Sweet substances: sucrose, glucose, aspartame
- Sour substances: citric acid, phosphoric acid
- Bitter substances: quinine, caffeine
- Umami substance: monosodium glutamate.

It is a myth that only certain areas of the tongue are sensitive to particular tastes. In fact, different areas of the tongue can be responsive to all the taste qualities; however, some areas are more sensitive than others.

2.1.3 Olfaction

Volatile molecules are sensed by olfactory receptors on the millions of hair-like cilia that cover the nasal epithelium (located in the roof of the nasal cavity). Consequently, for something to have an odour or aroma, volatile molecules must be transported in air to the nose. Volatile molecules enter the nose orthonasally during breathing/sniffing, or retronasally via the back of the throat during eating. There are around 17,000 different volatile compounds. A particular odour may be made up of several volatile compounds, but sometimes particular volatiles (character-impact compounds) can be associated with a particular smell, e.g. iso-amyl acetate and banana/pear drops. Individuals may perceive and/or describe single compounds differently, e.g. hexenol can be described as grass, green, unripe. Similarly, an odour quality may be perceived and/or described in different compounds, e.g. minty is used to describe both menthol and carvone.

2.1.4 Touch (somesthesis, kinesthesis and chemesthesis)

Somesthesis: The skin, including the lips, tongue and surfaces of the oral cavity, contains many different tactile receptors that can detect sensations related to contact/touch, e.g. force, particle size and heat.

Kinesthesis: Nerve fibres in the muscles, tendons and joints sense tension and relaxation in the muscles, allowing the perception of attributes such as heaviness and hardness.

Chemesthesis: Some chemical substances can stimulate the trigeminal nerves situated in the skin, mouth and nose to give hot, burning, tingling, cooling or astringent sensations, e.g. piperine in pepper, capsaicin in chilli pepper, carbon dioxide in fizzy drinks, coolants in showers gel, warming compounds in muscle rubs and tannins in wine. When sensed in the oral cavity, they form part of what are collectively known as mouth-feel attributes.

Texture perception is complex. Attributes of food texture can be divided into three categories: (i) mechanical, e.g. hardness and chewiness; (ii) geometric, e.g. graininess and crumbliness and (iii) mouth-feel, e.g. oiliness and moistness. These are generally described as being perceived during three phases: Initial phase (first bite), masticatory phase (chewing) and residual phase (after swallowing).

2.1.5 Audition

Sound is sensed by millions of tiny hair cells in the ear that are stimulated by the vibration of air from sound waves. The noise created when touching or stroking objects, e.g. fabric, gives an indication of texture. The noise emitted by food during eating contributes to the perceived texture of a food, e.g. crispness of an apple and fizz of a carbonated drink. When consumers eat food products, the sound waves produced can be conducted by the air and/or bones in the jaw and skull. The latter is known as intra-oral perception.

2.1.6 Multimodal perception

Although distinct sensory organs exist for each of the different senses, it is important to note that information from each of the sensory organs is often integrated in the brain. For example, the perception of flavour results from the interaction between taste, aroma, texture, appearance and sound. Sound can also affect the perception of touch. Similarly, texture perception is a combination of the visual, tactile and chemesthetic properties of the food or object under observation. The sensory professional should, therefore, be aware of how changes in one sensory property can affect others.

2.2 Factors affecting sensory measurements

Unlike instruments, human judgements can easily be affected by psychological or physiological factors. The sensory professional must be aware of these factors and ensure that the chosen procedure and experimental design eliminate or reduce such bias. This section highlights potential sources of error and suggests some strategies for reducing their effects.

2.2.1 Psychological factors

2.2.1.1 Expectation error

Knowledge of experimental objectives, or the samples to be evaluated, can influence an assessor's judgement. People tend to find what they expect to find. For example, codes such as 'A', '1' or round numbers (e.g. 100, 250) can be associated with a higher score. Other numbers can have particular associations, e.g. 999 or 911 and danger.
- ✖ Do not include people with product knowledge on the panel.
- ✔ Provide assessors with the minimum amount of information required to perform the test.
- ✖ Do not disclose information regarding the samples unless it is necessary for ethical procedures, e.g. use of novel ingredients.
- ✔ Code samples. Use codes such as random three-digit numbers and not letters or colours.

Sensory perception 7

2.2.1.2 Suggestion effect
Comments or noises made out loud, e.g. urghh! or Mmmm! can influence sensory judgements.
- ✔ Isolate assessors during sample evaluation, e.g. use of sensory booths.
- ✔ Discourage assessors from discussing samples before or after evaluation unless instructed to do so.

2.2.1.3 Distraction error
Assessors can be easily distracted from the task in hand, either by stimuli in the test environment, e.g. radios and other conversations, or by personal preoccupations, e.g. time pressure or domestic issues.
- ✔ Ensure test area is quiet.
- ✔ Create an environment that encourages professionalism amongst the assessors.
- ✘ Prohibit the use of electronic devices, e.g. mobile phones during testing.

2.2.1.4 Stimulus and logical error
Stimulus error occurs when assessors use additional information to make a judgement about the samples under assessment. When this stimulus is also logically associated with one or more of the characteristics under evaluation, it is called logical error. Some obvious examples are when products of a deeper colour or larger size are presumed to be more flavour intense, or when thinner skin creams are viewed as poorer quality. There are also other less obvious stimuli that may be exploited by assessors, such as cues regarding product branding; running a panel at an unusual time, which may prompt assessors to think there is a production problem; using more luxurious containers may lead assessors to think products are of higher quality.
- ✔ Ensure sample characteristics are consistent and/or mask irrelevant differences where possible, e.g. use of coloured lighting, blindfolds, nose clips and ear defenders where appropriate.

2.2.1.5 Halo effect and proximity error
Judgements concerning the rating of one attribute may influence the ratings of other attributes when assessors are asked to judge several attributes at once. This is more likely with untrained assessors. For example, a sweeter sample may be rated as softer, or stickier, than it would have, had these ratings been made on separate occasions. Furthermore, when rating several attributes at a time, the ratings of attributes following on from one another tend to be related.

8 Sensory evaluation

- ✔ Where possible, evaluate one, or at least a limited number of attributes, at a time.
- ✔ Where possible and appropriate, use trained assessors.
- ✔ Where appropriate, randomise the order of attribute evaluation if several attributes have to be rated at once.

2.2.1.6 Attribute dumping

If assessors are not given the opportunity to rate all the attributes they perceive as changing in the products under evaluation, they may still record this observation using the scales available. For example, if products are changing in terms of sweetness but no sweetness scale exists, they may register these changes on a flavour intensity scale such as strawberry flavour. This is known as 'attribute dumping'.

- ✔ Enable assessors to score all attributes which vary or indicate that opportunities to rate all varying attributes will be given.

2.2.1.7 Habituation

When assessors score similar products on a regular basis, e.g. on quality panels, they can develop a habit of assigning similar scores each time rather than scores which truly represent the samples.

- ✔ Vary products or introduce spiked samples from time to time.

2.2.1.8 Order effect

The score assigned to a sample can be influenced by the sensory character of the preceding product. For example, a sample may be rated as less sweet if it follows one of greater intensity. In addition, some sample positions are often favoured, e.g. products in position one are often scored higher in hedonic tests.

- ✔ Randomise or balance the order of presentation of samples (MacFie et al. 1989).
- ✔ For affective tests (see Section 5.4), use a dummy sample in position one.

2.2.1.9 Contrast and convergence effects

If two products in the sample set are strikingly different, assessors may exaggerate their ratings of this difference (contrast). If similar products are rated as part of a widely varying sample set, then the difference between them may be rated smaller than it actually is (convergence).

- ✔ Randomise or balance the order of presentation of samples.
- ✔ Consider removing outlying samples from the sample set.

2.2.1.10 Central tendency error
When using scales, assessors tend to avoid the extremes and confine their ratings to the middle of the scale. This is more likely to occur with untrained assessors or when assessors are not familiar with the product range.
- ✔ Train assessors in the use of the scale and expose them to a wide product range where possible.
- ✔ Use a large enough scale to differentiate between the products, particularly with untrained assessors.

2.2.1.11 Motivation error
A motivated panellist will learn better and, ultimately, perform more reliably. If assessors do not respect the panel leader or product manufacturer, they may rate samples based on how they feel. This can be an issue when using employee panels.
- ✔ Respect assessors.
- ✔ Give regular feedback to assessors.
- ✔ Carry out sessions in a professional manner.

Further information can be found on motivation in Section 4.5.5.

2.2.2 Physiological factors
2.2.2.1 Adaptation
Continued exposure to a stimulus results in a decrease in sensitivity to that stimulus and/or a change in sensitivity to other stimuli. Consequently, assessments of attribute intensity vary depending on the level to which the assessor has adapted to a stimulus. These are known as carry-over effects.
- ✔ Limit the number of samples presented.
- ✔ Ensure appropriate time intervals between samples to allow the sensory system to recover; this can be a matter of seconds, minutes or hours, depending on the stimulus, e.g. 'cooling' can take 10 minutes to recede.
- ✔ Ensure assessors take adequate breaks between single and sets of samples; the length of break will vary dependent on sample and test type.
- ✔ Provide assessors with appropriate palate cleansers, which ensure removal of any sample lingering in the oral cavity, e.g. milk rather than water may be needed for some spicy compounds.

2.2.2.2 Perceptual interactions between stimuli
Certain stimuli can interact to cause the following:
- *Enhancement (potentiation)*: The presence of one substance increases the perceived intensity of another, e.g. salt increases perceived intensity of chicken flavour.

- *Synergy*: The intensity of a mixture is greater than the intensity of the sum of the individual components, e.g. sweetness and sourness impact on strawberry flavour.
- *Suppression*: The presence of one substance decreases the perceived intensity of another, e.g. sourness reduces peach flavour.
- ✔ Where appropriate, employ careful experimental design to ensure that the effects of combined and individual stimuli are understood.

2.2.2.3 Physical condition

Health and nutritional disorders, together with the drugs prescribed to treat them, can affect sensory performance. Age and stress can also impact on sensory acuity, as can the time of day.

- ✔ Screen assessors prior to testing or remove assessor data if medical conditions or associated drugs affect the sensory performance.
- ✔ Instruct assessors to refrain from eating for at least an hour before sensory sessions.
- ✔ Schedule sessions for around a similar time each day – preferably between 10 and lunch.
- ✔ Monitor assessor's performance to highlight changes in sensory ability that may occur due to variation in physical state, e.g. age, hormonal state mood, etc.

2.2.3 Cultural factors

When working with assessors from different cultures or geographical location, the sensory professional needs to be aware of the impact that cultural effects can have on sensory data. For example, in some cultures, particular product codes may have significant connotations; eating in public may be considered as a social taboo; spiritual restrictions may impact on sample selection; group feedback may not be deemed acceptable. In addition, literal translations of questions and scale terminology may result in loss or change of meaning. The use of a scale can vary across cultures, e.g. some tend to score much higher or lower than 'average' when using the hedonic scale.

- ✔ Be sensitive to coding issues.
- ✔ Clarify translations of sensory scales or questionnaires into other languages, e.g. the use of back translation.
- ✔ Be aware of cultural tendencies – these will have an impact on many aspects of sensory testing such as products, protocols, scale use and feedback.
- ✔ Build up information on cultural norms from different cultures or countries.

3 Planning your sensory project

3.1 Setting objectives

It is vital to understand the objectives of a project as these are key factors in determining the test type and, consequently, the experimental design and statistical analysis required to meet these objectives. Commonly asked questions such as 'are these samples different?', 'which is the preferred sample?', 'how do these samples compare to the competition in terms of texture?' and 'what is the optimum cooking temperature to create the most acceptable golden colour?' would all require different sensory methods and experimental designs. Often, a client will want to know the answer to all of these questions; however, time and financial constraints may require that the objectives are prioritised. It is imperative that the potential outcomes of different methodologies be highlighted in advance so that clients are aware of any limitations, e.g. a discrimination test identifying the sweetest sample will not allow any conclusions to be drawn about preference or about how sweet it actually is. When working with internal or external clients, the objectives for a project should be documented alongside all other pertinent information (see Chapter 6).

3.2 Product type

When selecting an appropriate methodology to satisfy the desired objectives, it is important to consider the product type as this may have a serious impact on test design. In some instances, products may need to be tested in combination with other foods, e.g. breakfast cereal may be presented with milk, or olive oil may be presented with a neutral carrier such as bread. When the test objective includes some aspects of performance, products may need to be tested in the context of their use, e.g. shampoo, safety razors and skin creams. In this instance, careful consideration needs to be made to other aspects of experimental design (see Section 3.7). For example, some samples have intense carry-over and need to be presented monadically with large breaks in between.

3.3 Budget

Financial constraints have to be considered in any test design. In some instances, the cost associated with the 'ideal' design exceeds the budget and appropriate compromises such as reducing the number of products, assessors or replicates are necessary. It is important to understand the consequence of every compromise with regard to the quality of the data and the conclusions that can be drawn.

Reducing the number of assessors from the maximum to the minimum recommended would be acceptable, whereas reducing the number even further may have a deleterious effect on the power of the statistical test (see Section 5.2.4.1) and, therefore, on the likelihood of reporting significant differences between samples. Furthermore, smaller groups of assessors are less representative of their population.

Reducing the number of samples can be an effective way of cutting costs; however, removing replication from the design of certain sensory methodologies, e.g. profiling, can be very dangerous. Furthermore, some methodologies, e.g. preference mapping, require a minimum number of samples and reducing the number below this would render the test invalid.

3.4 Timings

When designing a sensory test, a deadline may affect the decision over which methodology to use. It is important to know in advance if any deadlines exist. In studies that require several test elements, e.g. consumer tests, descriptive analysis, instrumental analysis, shelf life testing, the co-ordination of these elements is crucial for samples whose sensory properties change with time.

3.5 Selecting the test method

There are many sensory tests and a multitude of different situations in which they can be applied. The test employed will depend on the test objective(s). It is imperative that the specific objective of any sensory test is probed and clarified before testing begins.

Often, a series of tests is required to meet the objectives. Careful consideration needs to be given to the order in which different tests are performed. It is futile to carry out large consumer trials concerning the preference for two products without prior sensory information as to whether a significant perceivable difference exists between them.

The most appropriate test may not be the most cost-effective or feasible with the amount of sample or assessors available, and consequently, some form of compromise may need to be reached.

Details of different test methods and their application are given in Chapter 5.

3.6 Setting action standards

Action standards are the criteria that must be met to take a course of action based on test results. They should be set in advance of the test being carried out. Factors to consider include size of the opportunity; business risk and stage of testing, which will determine how strict the criteria are, as well as product type, new or existing product category and communication, e.g. whether a product improvement will be flagged to consumers or not. Action standards may include number and type of consumer, statistical criteria, and elements of the test design (are the products going to be presented simultaneously or sequentially? will the products be branded or unbranded? will the products be presented with a marketing concept?).

A simple example of an action standard to decide whether an optimised product should be substituted for the current product is as follows.

> **For the optimised product vs. the current project: a ratio of 55:45 in preference in an unbranded product test with heavy product users in the target user group; and at least parity in preference in an unbranded product test with product category users.**

and

> **For optimised product vs. competitor product: at least parity in preference with product category users.**

A simple example of an action standard for cost reduction is as follows.

> **At least parity, in an unbranded preference test, for a cost-reduced product over current product with heavy product users.**

Careful consideration needs to be given to the **appropriateness** of action standards. For example, in the latter example on cost reduction, an action standard of 'no difference in a similarity test with heavy users' may be unnecessarily strict and even unachievable, thus wasting time and resources and delaying the decision to substitute a lower cost product.

3.7 Experimental design

It is important to design an experiment that enables the test objectives to be met. Too often, there is a rush to move quickly to the testing stage without enough consideration for the design, or a temptation to carry out some *ad hoc* testing without thinking the design through. A lack of planning can render the experiment useless, as crucial elements, such as controls or replication, may be omitted.

The statistical analysis to be used needs to be taken into account when designing the experiment, in order to include the elements necessary in the design that will enable those analyses to be applied. It is also important to consider the broader context, so that design features that allow studies to be linked where necessary can be incorporated.

3.7.1 Treatment structure

This describes the experimental treatments that have been applied to the study samples. At its simplest, the treatment structure may have only one level, for example different commercial samples selected for comparison, or a development sample compared to existing lines. In this case, the samples are not related to one another except as different examples of the same product type.

A more complex treatment structure could have varying levels of two treatments. For example, a range of four commercial apple pies, assessed heated and cold. In this instance, one treatment would be the four different apple pies and the other would be the two serving temperatures. This structure would create eight samples for assessment.

Alternatively, the samples may have been created from an even more complex treatment structure in which one or more different processes or ingredient levels (design factors) are varied systematically to produce a range of samples that allow the effect of each design factor to be determined individually and in combination with one another (interaction). In this instance, more specialist designs, e.g. factorial designs, fractional factorial designs and response surface designs, should be used (*see Eriksson et al. 2000 for further information on different types of design structures*).

3.7.1.1 Control samples
Wherever possible a control sample should be used, so that comparisons can be made with treated samples to determine if the treatment has an

Planning your sensory project 15

effect. This is good general practice in experimental design but is especially important in sensory evaluation, as assessors are better at making relative judgements than absolute judgements, so that a comparison with control is key. Some tests already include a control as an inherent part of the design, e.g. in same–different tests the control is the same pairs. If this is not the case, careful consideration needs to be given to the control to deliver the desired information and sometimes more than one control may be necessary.

The control may be a baseline sample or a blank sample – a sample that has not had the treatment under investigation applied, such as pure diluent in a concentration series. It is important that control samples remain consistent across their use. When storing control samples, care should be taken to ensure they remain stable. Controls should be easy to produce consistently. Multiple batches may be required, as the control is often used over a longer period of time than treatment samples and/or across multiple studies to enable the studies to be compared. New batches of control should be checked to ensure their sensory and physicochemical properties are consistent with previous batches.

3.7.2 Design structure
3.7.2.1 Randomisation
When multiple samples are presented to assessors, it is important to randomise the presentation such that each assessor receives samples in a different order. This reduces systematic carry-over and order effects. There are circumstances, however, where randomisation is not possible, e.g. when only one sample can be prepared at a time and must be served immediately to all assessors. In this instance, a dummy sample will help reduce first order effects but some error is unavoidable. Moreover, there are instances where randomisation is not appropriate, e.g. where assessors are being trained or screened and their results are to be compared directly. In this instance, randomising sample presentation will introduce order effects to comparisons of assessor's results. In practice, simple randomisation of sample presentation is rarely used in favour of a balanced test design.

Note: Randomisation is also critical in the allocation of sample sets to the assessors; therefore, where several presentation orders have been created for samples, they should be randomly assigned to the different assessors. In practice, this tends to happen automatically without conscious decision. When describing a design as 'randomised' and 'balanced' (see the following section), randomised here refers to the allocation of sample sets to assessors.

3.7.2.2 Balanced test designs

A further reduction in order effects can be achieved by balancing the presentation order such that each sample is given in every presentation order an equal number of times. This type of design is known as a *Latin Square*. Many specialised versions exist, e.g. Williams Latin Square, whereby each sample occurs in every presentation order and also before/after every other sample in the design, an equal number of times (see Appendix 1 for examples of Latin Squares and Williams Latin Squares).

3.7.2.3 Complete and incomplete block designs

An experimental design can be divided into subsets known as blocks; these can be either samples, assessors or other design factors (treatment structure). For example, an assessor is a block that contains all the observations made by that assessor; a sample is a block that contains observations made by all assessors.

Identifying different blocks in an experiment allows variation within and between them to be analysed.

A complete block design is considered to be the most ideal, whereby, all samples are presented to each assessor during one session. When factors such as large sample numbers or strong carry-over prevent this, an incomplete block design must be used. In this instance, all samples may be presented over several sessions, or only a subset of the total number of samples is presented. The latter is commonly used for screening large numbers of samples or in consumer assessments when it is not possible to ask the respondents to return. Specialised data analysis, e.g. analysis of variance (ANOVA) for balanced incomplete block (BIB) designs, is required.

3.7.2.4 Common sensory designs

Completely randomised design

In the completely randomised design (CRD), products are assigned randomly to assessors who assess only one product each. Typically, several results are collected for each product and every product is seen by a different group of assessors. This design is not ideal as assessors cannot be identified as separate 'blocks' and, therefore, variation in their results cannot be considered in the data analysis.

Randomised complete block design

Randomised complete block design (RCBD) is the most commonly used design. Here, all products are assessed by all assessors. Products are presented in a randomised balanced order of presentation across the panel. Assessors may also make replicate judgements for each product over one or more sessions. Variation from individual assessors and the replication can be considered in the data analysis. In behavioural sciences, this type of design is also called a 'repeated measures' design.

Balanced incomplete block
In the balanced incomplete block (BIB) design, a subset of the total number of products is presented during a session. Ultimately, each assessor may see all samples over several sessions (although in practice, it has been common to present samples over several sessions and call this a 'complete' design), or only a subset of the total group. In either case, samples presented within a session are randomly allocated and their order is balanced such that each assessor receives the same number of samples, each sample is seen an equal number of times across the session and each sample is seen in combination with every other sample an equal number of times across the session. Software programs are used to help design BIBs.

3.7.2.5 Sample presentation techniques
In addition to the overall experimental design and presentation order of samples, the sensory methodology may dictate the technique used to present samples to assessors. The following are the most commonly used.

Monadic
A single sample is presented for assessment. This technique produces data that are free from sample interaction effects. It can be used to provide diagnostic information, and develop norms and action standards.

Sequential monadic
The samples are presented individually in a series, one after another, for assessment. This is the most appropriate style for methods such as rating attribute intensity in descriptive analysis or rating acceptability in affective tests, as samples are rated independently and not directly compared to one another during the sensory assessment.

Comparative or simultaneous designs
The samples are presented together. This is a requirement for techniques requiring comparative judgements, such as ranking, triangle tests, and paired comparison.

Proto monadic
Samples are presented as pairs; the first sample is rated monadically and the second one is compared directly to the first. For example, the first

sample is rated for liking, and the second one is compared to the first in a paired comparison that asks 'which is the preferred sample?'.

3.7.2.6 Replication

Replication is the assessment of a sample by an assessor on more than one occasion. It increases the reliability and statistical power of a test and, hence, the likelihood of finding a difference. Variability often decreases and, hence, performance often increases upon replication.

It is essential to carry out replication in descriptive analysis. The number of replications will depend on the test objectives. For example, fewer replications are likely to be needed in the earlier stages of a project when less important decisions are being taken, e.g. screening. If the results of the test are more important, e.g. final product selection or building a model, then it is prudent to use more replicates. Stone and Sidel (2004) suggest that four replications are optimal. The number of replications achievable in practice may be limited by the amount of time available, the amount of product available and the number of products to be assessed. Increasing the number of assessors is sometimes used to offset the number of replications, but this can be a risky strategy as it does not compensate for assessors who give idiosyncratic judgements and skew results.

Replication is sometimes used in discrimination testing. The practice of using replicate judgements to boost overall assessor numbers is not recommended.

Replication is rarely used in affective testing, as it is more time consuming and costly, and adds to complexity if consumers need to attend for multiple sessions.

For both discrimination and descriptive testing, replication can be used during assessor selection to ensure they are consistent and reliable. Replication can also be used to determine assessor and panel reliability. It can be used to identify problems occurring in the test design, e.g. excessive variability of samples, fatigue, carry-over and learning.

True experimental (treatment) replicates are separate batches of the same sample – this allows information to be concluded about the samples. Assessor replicates can be made on experimental replicates. The latter are seen as separate samples in the study.

3.7.2.7 Panel size

The total number of assessors is directed by the test objective and methodology. Many international standards and guidelines exist to direct the sensory scientist to an appropriate sample size; these are referred to under the relevant methods in Chapter 5.

Planning your sensory project 19

The number of assessors affects the power of the statistical tests used to analyse data. Consequently, it is advantageous to maximise the number of assessors to improve the statistical discrimination between the products. It is not, however, acceptable to ask assessors to replicate their judgements and treat the data as if it were produced by different individuals.

3.7.3 Dos and don'ts

- ✔ Use an appropriate experimental design to meet test objectives.
- ✘ Don't test *ad hoc* samples to 'see what happens'. Use a formal experimental design to systematically assess effects of treatments – it will give more information.
- ✔ Include a control, baseline or blank sample where possible.
- ✔ Use replication where appropriate, e.g. for descriptive analysis.
- ✔ Use sufficient numbers of assessors to meet the statistical power requirements of the sensory test.
- ✘ Do not present samples in the same order to all assessors. Use a randomised or balanced serving order.
- ✔ Check the serving order design to ensure it is well balanced; particularly, check that each sample is seen in combination with every other sample an equal number of times.
- ✘ Do not assume that a random serving order will be well balanced. Check the design.

3.8 Data analysis

Sensory data analysis is carried out using a specialised field of statistics called *sensometrics*. Many areas in sensometrics are complex, and it is often prudent to consult a statistician for advice on data analysis.

It is important to determine the statistical analysis to be applied during planning before the data have been collected, particularly as the choice of analysis may influence the elements of the experimental design. The analysis used must enable the test objectives to be met and should be focused on answering questions pertinent to meeting test objectives.

This section provides a general background to data analysis and its application to sensory analysis. Data analysis for specific tests is covered in Chapter 5.

3.8.1 Types of data

Sensory methods and/or scales produce data with different properties. It is these properties that are exploited in data analysis and, therefore, it is important to marry the type of analysis with the correct data type.

20 Sensory evaluation

Nominal data: These represent different groups or categories, e.g. different types of fruits. Numbers can be used as labels but do not carry any numerical value. Discrimination tests yield nominal data.

Ordinal data: Categories (numbers) on an ordinal scale represent increasing or decreasing magnitude of a specified attribute. However, a key feature of an ordinal scale is that the relative distance between the categories (numbers) is not always equal. This is a crucial point as it has important consequences for the choice of data analysis (see Section 3.8.2.2). Ranking and some rating scales yield ordinal data.

Interval data: The categories (numbers) on an interval scale are equally spaced and have true numerical value. Interval scales, however, do not have a true zero, e.g. 'degrees centigrade'. Consequently, although the difference between 10°C and 20°C is the same as between 20°C and 30°C, 40°C is not twice as hot as 20°C.

Ratio data: The categories (numbers) on a ratio scale have numerical value and a true zero, e.g. 'weight'. The difference between 10 and 20 kg is the same as between 20 and 30 kg, AND 40 kg is twice as heavy as 20 kg.

Descriptive rating scales used by trained assessors yield interval or ratio data.

3.8.2 Distribution of data

The raw data can be visualised by plotting 'the number of times a response is given' against 'the magnitude of the response', e.g. the number of people rating a sample for acceptability against each score that can be given for acceptability. This type of plot is known as a *frequency distribution*.

3.8.2.1 Normal distribution

The normal, or Gaussian, distribution is a bell-shaped symmetrical curve, which can vary in height and width, and is the most commonly occurring distribution in nature.

The mean, median and mode coincide and the distribution has many interesting properties which form the basis of much statistical theory, and are exploited in parametric statistical tests. Data from interval and ratio scales can produce normally distributed data, although this should be confirmed before further analysis is performed; many statistical packages offer tests that determine normality, e.g. Kolmogorov–Smirnov test.

3.8.2.2 Nonnormal distributions

When the frequency distribution does not follow this common shape or share the properties of normal distribution, e.g. skewed distributions, it is classified as a nonnormal distribution. In this instance, parametric

Planning your sensory project 21

tests are not appropriate and nonparametric tests are used instead. Data from nominal and ordinal scales, e.g. discrimination and ranking tests, produce nonnormally distributed data. Furthermore, data from interval or ratio scales may not turn out to be normally distributed; in this instance nonparametric tests should be applied.

3.8.2.3 Nonparametric tests
This is the name given to the group of inferential statistics that can be applied to nonnormally and normally distributed data, e.g. binomial tests, Friedman's ANOVA for ranked data and chi-squared test.

3.8.2.4 Parametric tests
This is the name given to the group of inferential statistics that can be applied to normally distributed data, e.g. students t-test and ANOVA.

3.8.3 Samples and population
The aim of experimentation is to draw conclusions about a population. The 'population' may be a product or a group of consumers such as potential buyers of a product. Usually, it is not possible to test the entire population due to cost and time constraints. Therefore, a sample (or samples) is taken and inferences about the population are made from that sample.

It is vital that the sample is a representative of the population in order to draw correct conclusions. The larger the sample, the more representative it is. Samples are often collected randomly, but care needs to be taken to ensure sampling is truly random. Sampling needs to be undertaken in an unbiased fashion. For example, when sampling a manufactured product, the sample may need to consist of different batches. When sampling fresh fruits and vegetables, the sample may need to be taken from different plants in different areas and fields. When sampling consumers, stratified sampling may be used to match the sample demographics to those of the target population, i.e. buyers, users, and so on.

The true mean, median, mode, etc., of the population are known as *parameters*. The estimates of these parameters derived from a sample are known as *statistics*. Note that conventionally roman letters are used to denote sample statistics and greek letters to denote population parameters.

Better information can be obtained about the population using more powerful statistical tests. There are two types of statistical techniques used to analyse data: descriptive and inferential statistics. Descriptive statistics summarise the data, e.g. mean scores. Inferential statistics allow conclusions to be made concerning the population, based on a sample,

22 Sensory evaluation

to a certain degree of confidence, e.g. determine if mean scores are significantly different.

3.8.4 Data handling

3.8.4.1 Data checks

It is vital that initial checks are made on any data set to ensure that the raw data are generally as expected and that no errors have been made in the data entry. The data should then be checked for outlying values and decisions taken as to whether they should be removed, replaced or left in. Outlying data points may still be representative of the sample and should only be removed if there is a strong rationale to do so, e.g. assessors are unwell or reported incorrect use of a scale, a batch of raw material was out of specification and so on. The removal of outliers should be reported in any final documentation. Note that the presence of outliers in the data may have consequences for the type of data analysis applied (see Section 3.8.5.1 and 3.8.7).

Any missing values should be checked first to ensure that they are truly missing. Some statistical tests cannot be executed with missing values. In these instances, several options are available. Where possible, missing values can be replaced by the mean, e.g. replacing a missing assessor's data point with the panel mean, or data concerning that product, or that assessor, can be removed. Note that replacing missing data with the mean or removing products/assessors from the data set will impact the statistical analysis. Data should be replaced or removed only with caution and any amendments should be considered during interpretation of the results.

Finally, checks should be made to ensure the data are in the correct format for any particular software being used for the analysis.

3.8.4.2 Data transformation

In some instances, data transformation may be required before statistical analyses can be applied to parametric data. Typical examples include when attribute data have been collected on different scales, when assessors have used scales differently or when data are not normally distributed.

Many types of data transformation exist. The most common transformations required for sensory data are normalising, standardising and logarithmic transformations.

Normalising: Data are amended such that the maximum value achieved by all the assessors (or products) is identical. Data can be normalised to either 100 or the maximum value achieved by any one assessor or sample with all other data calculated as a percentage. The purpose of normalising is to compare all assessors (or products) on the same scale.

Standardising: Data are amended such that all variables/attributes are measured on the same scale with equal variance. The purpose of standardisation is to ensure all variables/attributes are on equivalent scales such that no one attribute has more influence than another in a statistical analysis, such as principal component analysis (PCA), and ensure equal variance. The standardised data are obtained by subtracting the variable/attribute mean and dividing by the standard deviation.

Logarithmic transformation: The natural or base 10 logarithm of each data point is calculated. The purpose of a logarithmic transformation is to normalise the distribution of skewed data, linearise the relationship between variables and/or stabilise the variance. Data collected on ratio scales, e.g. magnitude estimation, will require a logarithmic transformation prior to further statistical analysis.

3.8.5 Descriptive statistics

It is common to summarise sensory data in terms of measures of central tendency (averages) and measures of dispersion (spread of data).

3.8.5.1 Central tendency

Measures of central tendency include means, medians and modes.

Mean: The arithmetic mean, $\bar{x}$, commonly known as the average is calculated as follows.

$$\bar{x} = \frac{\sum x}{n}$$

where Σx is the sum of all the observations and n is the number of observations. It should be used only with parametric data that are normally distributed, as it is influenced by outliers and may give misleading results. The geometric mean is also used in sensory evaluation, e.g. for serial dilutions, when it is appropriate to take logarithms of values, and when analysing ratio data.

For the data set 1 1 2 2 2 2 4 5 6 6 7 8 9 9 9, the mean is **4.9** and the geometric mean is **3.8**.

Median and modal averages should be used when data are not normally distributed or is collected on nonparametric scales.

Median: This is the middle value of a set of values arranged in order. For data sets with an even number of data, the two middle data points are averaged.

For the data set 1 1 2 2 2 2 4 5 6 6 7 8 9 9 9, the median is **5**.

24 Sensory evaluation

For the data set 1 2 5 6 10 25, the median is **5.5**.

Mode: This is the most frequently occurring value. It is possible for a data set to have no mode if no value occurs more frequently than any other, or more than one mode, e.g. a bimodal distribution has two modes.

For the data set 1 1 2 2 2 2 4 5 6 6 7 8 9 9 9, the mode is **2**.

Note that confidence intervals, e.g. 95% confidence intervals, can be used to infer the range within which the expected population mean (or median) value would be expected to lie 95% of the time (see O'Mahony (1986) for further information on calculation of confidence intervals).

3.8.5.2 Dispersion

Measure of dispersion gives an indication of the variability (spread) of the data around the average. Typical measures include standard deviation, variance and percentile ranges.

Standard deviation (s): This is a measure of dispersion used alongside mean values, expressed in the same units of the original scale. It is calculated by taking the square root of the sum of all the scores squared minus the sum of the scores, squared, divided by the number of observations and dividing it by the square root of the number of observations minus 1:

$$s = \frac{\sqrt{\sum x^2 - \left(\sum x\right)^2 / n}}{\sqrt{n-1}}$$

For the data set 1 1 2 2 2 2 4 5 6 6 7 8 9 9 9, the standard deviation is **3.1**.

Note that in some statistical tests another measure known as the 'variance' is used. This is simply the standard deviation squared.

Standard error (S): The standard error is a 'population' parameter. It is the theoretical standard deviation of the whole population. If an experiment is repeated several times, several means could be obtained and the standard deviation of the samples' means could be calculated. This is rarely possible and seldom performed due to time and cost constraints. Fortunately, the standard error can be derived from the standard deviation of one sample. It is calculated by dividing the standard deviation of the sample by the square root of the number of observations.

$$S = \frac{s}{\sqrt{n}}$$

Planning your sensory project 25

For the data set 1 1 2 2 2 2 4 5 6 6 7 8 9 9 9, the standard error is **0.79**.

Percentile ranges: In cases where data are not normally distributed or collected on nonparametric scales, quartile ranges can provide an indication of the dispersion of the data. Quartile ranges are obtained by ordering the data and dividing them into subsets such that a certain percentage is below or above. These are typically the 25th, 50th and 75th percentiles.

3.8.6 Inferential statistics

3.8.6.1 Hypothesis testing

The identification of a test objective, and the subsequent selection of a test method, automatically results in the identification of a hypothesis to test. For example, if the test objective is to determine whether two samples are perceived to be different, the method selected would be testing the hypothesis that the samples may or may not be different. In statistical terms, 'hypothesis testing' is a more formal procedure in which a *true* statement (null Hypothesis) is made together with an alternative statement (or research hypothesis). Commonly, the null hypothesis states that samples are the same, whereas the alternative hypothesis states that they are different. A statistical test performed on the data then determines whether or not there is sufficient evidence to reject the null in favour of the alternative hypothesis.

Significance level: type I and type II errors

In statistical tests, it is important to minimise the risk of making an incorrect decision. There are two types of incorrect conclusions that can occur; these are known as type I and type II errors.

Type I error

This is the risk of rejecting the null hypothesis when it is correct (true), i.e. the risk of saying that the samples are different when, in fact, they are not. This is minimised by reducing the significance level, or α risk of the test. The most commonly used significance level is 5%; this means that 5 times out of 100, samples will be stated as different when they are not.

Type II error

This is the risk of not rejecting the null hypothesis when it is incorrect (false), i.e. the risk of saying that the samples are the same when, in fact, they are different. This is minimised by reducing the β risk of the test.

Critical value and *p*-value approach

For each statistical analysis, a test statistic is calculated from the raw data; this is used to determine the outcome and draw conclusions. The test statistic is compared to a critical value which represents the value of the test statistic calculated for the significance level of the test. If the test statistic exceeds the critical value, the null hypothesis is rejected and the samples are considered to be significantly different. If, however, the test statistic does not exceed the critical value, the null hypothesis is not rejected and there is considered to be insufficient evidence to determine a significant difference between samples.

Alternatively, the probability of making a type I error can be calculated for the test statistic. If the probability, or p-value, is less than the significance level of the test, the null hypothesis is rejected and the samples are considered to be significantly different. A p-value of less than 5% (0.05) represents even less risk of making a type I error. If, however, the p-value is greater than the significance level of the test, the null hypothesis is not rejected as there is an even greater risk of doing so in error.

Traditionally, the critical value comparison was the most common means of determining the outcome of the hypothesis test. The development of computer software packages for analysing data has resulted in the critical values and/or the p-values being displayed.

3.8.7 Choosing the appropriate statistical test

The choice of statistical test used to analyse sensory data should not be an afterthought. It is an integral part of project planning which must be considered alongside the objectives and experimental design (Figure 3.1).

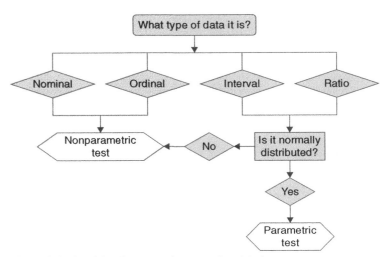

Figure 3.1 Identifying the appropriate type of statistical test.

Planning your sensory project 27

In order to identify appropriate statistical tests for sensory data, the following questions should be considered.

3.8.7.1 What type of data are they?
This will determine the type of test that can be carried out. If the data are interval or ratio data, then a parametric test is appropriate; however, if the data are not normally distributed then a nonparametric test should be used. Parametric tests assume that data are normally distributed. If the data are nominal or ordinal, then a nonparametric test should be used. Nonparametric tests make no assumptions about the distribution of the data.

Note: The data from discrimination tests are nominal – in fact binomial. Consequently, the data are analysed using binomial statistics, or, in the case of a same–different test, chi-square analysis.

3.8.7.2 What is the test objective?
Generally, sensory scientists are exploring whether significant differences exist between samples/assessors/consumers, and/or if significant relationships exist between variables, e.g. sweetness and liking, flavour intensity and fat content.

3.8.7.3 How many samples are there?
Some statistical tests are specifically designed for comparisons between two samples, whereas others are designed for three or more samples. It is not appropriate to apply tests designed to compare two samples to pairs of samples in a sample set, unless the objective is to compare each product with one control.

3.8.7.4 Are samples related (paired/dependent)?
In sensory testing, if samples are evaluated by the same assessors, e.g. as in quantitative descriptive analysis, they are described as 'related' (referred to as 'paired' or 'dependent' in some text books). If samples are evaluated by different groups of assessors, e.g. if two products have been profiled by two different panels in different countries, they are described as unrelated or unpaired/independent. Some statistical tests are designed for related samples and some for unrelated samples.

3.8.7.5 Selecting the test
Once the answers to the earlier questions are ascertained, it is possible to identify the appropriate statistical test for the data. The flow charts shown

28 Sensory evaluation

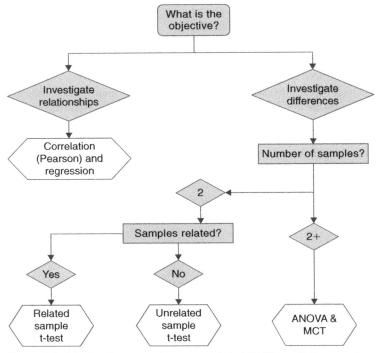

Figure 3.2 Selecting the appropriate parametric test (MCT – multiple comparison test).

in Figures 3.2 and 3.3 help to identify common statistical tests used for sensory data. Specific examples of common statistical tests can be found alongside examples of sensory test methods in Chapter 5. *For more detailed information on statistical tests see Ashcroft and Pereira (2003), Dijksterhuis (2008), Meilgaard et al. (2007), Meullenet et al. (2007), Naes and Risvik (1996), O'Mahony (1986) and Rayner et al. (2006).*

3.8.8 Dos and don'ts
- ✔ Plan the statistical analysis before starting the experiment. The experimental design may need to be modified to enable the selected analysis to be carried out.
- ✖ Don't throw everything at the data – selectively apply statistical analyses that will deliver the results to meet objectives.

Planning your sensory project 29

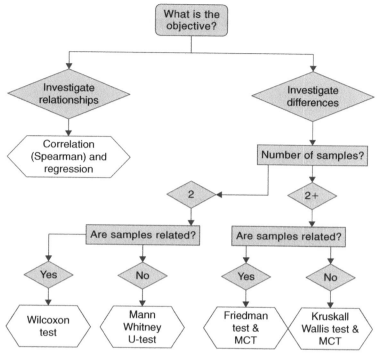

Figure 3.3 Selecting the appropriate nonparametric test (MCT – multiple comparison test).

- ✔ Apply appropriate statistical analysis techniques to the data. Avoid classic mistakes, for example, using means when it is not appropriate for the data. Skewed data requires geometric means and bimodal distributions require graphical description rather than central tendency analysis.
- ✔ Draw the correct conclusions from the data. A typical error is to assume that a sensory difference has been 'caused' by a particular factor when no evidence is present.
- ✔ If in doubt, consult a statistician.
- ✘ Do not expect one test to answer all questions. In many situations, a series of tests is needed.

4 Requirements for sensory testing

4.1 Professional conduct in sensory testing: health, safety, ethical and legal considerations

4.1.1 Importance

Sensory professionals have a duty of care to ensure the health, safety, ethical and legal treatment of assessors. Sensory testing has the potential to cause severe injury and even death, e.g. due to food poisoning or ingestion of toxic ingredients. It is, therefore, important that all aspects of testing are considered to ensure procedures and practices meet legal and ethical requirements. These include protection of assessors, safety of sample ingredients, production and preparation, and test protocol. On occasions, sensory professionals may be required to carry out work that goes beyond traditional sensory studies on locally marketed products such as nonapproved and novel ingredients, novel assessment protocols, claims support, efficacy assessment and safety assessment. In these situations, additional measures need to be adopted to preserve the safety and rights of volunteers, such as ethical committee review, specialised informed consent process and ongoing care for assessors. Ensuring safety, rigorous assessment procedures will also minimise legal risks and instil confidence in assessors that their well-being is of paramount importance.

Questions that need to be considered include the following:
- What is being measured?
- How is the measurement being made?
- Who is taking part in the research?
- Are the ingredients proven to be safe?
- Is safety information to be collected as well?
- Will these data be used to support claims?
- How will care for assessors be provided in the event of an adverse reaction?
- Does the trial present any moral and/or ethical issues?

Requirements for sensory testing 31

4.1.2 Legislation and professional codes of conduct

There are laws and codes of practice governing the testing of subjects. These include the following:
- Nuremberg Code (1949) of ethics in medical research.
- Declaration of Helsinki (World Medical Association 2004).
- European Directive 2001/20/EC (EU 2001) and 2005/28/EC (EU 2005) on good clinical practice in clinical trials.

A professional code of conduct for sensory professionals has been written by the Professional Food Sensory Group (PFSG) of the Institute of Food Science and Technology (IFST) in the United Kingdom (see Appendix 2 and www.ifst.org.uk).

Professional codes of conduct for market research have existed since 1948. The ICC/ESOMAR International Code of Marketing and Social Research (International Chamber of Commerce and the European Society for Opinion and Marketing Research 2007) covers general issues, rights of respondents, the professional responsibilities of researchers, the mutual rights and responsibilities of researchers and clients, and the implementation of the code. Guidelines also exist for different types of market research. Many market research societies in individual countries have used this as a basis for their own codes. There is also an international standard for market research: ISO 20252:2006.

Further information can be found on the websites for ESOMAR (www.esomar.org) and the Market Research Society (MRS) (www.mrs.org.uk).

General overarching ethical codes also exist for scientists, e.g. 'Rigour, respect and responsibility: a universal ethical code for scientists' (Department for Innovation, Universities and Skills 2007).

4.1.3 Protection of assessors

The following processes and procedures have been put in place to protect research subjects, including sensory assessors. Careful consideration should be given to whether they apply to each piece of research undertaken. If in doubt, a more cautious approach is advised.

4.1.4 Independent ethical committee review

Ethical clearance is required for trials in which the study poses any physical or psychological risks to the subject which are significantly greater than the risks of a desirable, high-quality everyday life. The following examples are taken from the UK Regulations governing ethical committees. They represent specific situations in which ethical approval must be sought. In some organisations, however, ethical approval might be

sought for a broader range of studies, e.g. those containing alcohol. If in doubt, the sensory professional should exercise caution.
- Any product containing nonapproved ingredients or processing needs to be tested. (These must have been assessed for safety first as described here.)
- Any unusual experimental procedure is to be used, that is assessors are required to do something that would not be considered a normal event during everyday life.
- Any trial involving children.

The Ethical Review Committee is an independent committee of people, e.g. teachers, doctors and so on, who determine if trials involving human subjects are ethical, based on the reasons for conducting the test, the protocol, safety information, information given to assessors, etc. The recommended committee composition is a minimum of five members, at least one lay member and at least one member independent of the trial site.

An Ethics Committee's function and operation are as follows:
- Work to written operating procedures.
- Maintain written records.
- Make decisions only at meetings with a quorum (composition of which is specified within committee operating procedures).
- Conduct initial and continuing review of trials.
- See and approve protocol changes before approval of the study. Protocol changes that are administrative in nature do not require review by the full committee. All other changes require review by the full committee.
- In the United Kingdom, ethics committees for Phase 1 clinical trials must be recognised by the United Kingdom Ethics Committee Authority (UKECA).

Generic approval may be granted for a test if all the ethical issues relating to a particular test have been approved previously, so that there is no need to apply for individual approval for that test. The Ethical Review Committee Secretariat is notified of the tests conducted under generic approval and these would be reviewed retrospectively by the committee to ensure that generic approval was appropriate. In most cases, ethical clearance for sensory testing is not necessary.

4.1.4.1 Informed consent

Informed consent must be collected from all assessors prior to starting a study to indicate they are fully informed of the nature of the experiment, samples they will be ingesting/using, any associated risks, that they can withdraw at any time, confidentiality, medical assistance in the event of adverse reaction and compensation. Information may be supplied verbally

and/or in written form and assessors should have the chance to ask questions. The following points apply:
- Voluntary consent is absolutely essential.
- Assessors should have the legal capacity to give consent.
- Assessors are able to exercise free power of choice without any element of fraud, deceit, duress, over-reaching or other form of coercion.
- Assessors should have sufficient knowledge and understanding of the subject matter to allow an informed decision.
- Assessors should be provided with sufficient time to read through and understand the information provided before providing their written consent and participation in the study.

General informed consent may be gained from sensory and consumer panels to cover work on typical market products in some countries. Additional consent for individual studies, however, may still be required for any unusual studies. Specialised consent forms must be used for studies that have been through ethical clearance.

Information to be supplied to the assessors includes the following:
- The nature, duration and purpose of the study.
- The method and means by which the study is to be conducted.
- The inconveniences and hazards to be reasonably expected.
- The effects upon health or person which may possibly come from participation in the study.

Additional requirements for informed consent include the following:
- Assessor must agree to access of the trial records by the employer/sponsor, auditor and regulatory authorities.
- Investigator must ensure that no document containing the assessor's full name leaves the trial unit so that confidentiality is not breached.
- Copy of the consent form should be given to the assessor.
- Any update should also be provided to the assessor.

4.1.4.2 Selection of assessors

Before assessors enter into the study, it is the responsibility of the researcher to ensure that it is safe and ethical for them to participate. The following points should be considered when selecting assessors to ensure their safety and ethical treatment.
- Assessors do not have an allergy or intolerance to the products in the study.
- Assessors do not have a medical condition or do not take medication that could cause an adverse reaction.
- Assessors are not asked to work on products they normally avoid on moral or religious grounds, e.g. vegetarians are not asked to eat meat.
- Assessors are not asked to work on products they dislike intensely.

34 Sensory evaluation

4.1.4.3 Health monitoring

It may be advisable to monitor any medical changes in assessors on long-standing panels. Procedures to do so may include the following:

- Upon selection, make a record of assessors' state of health, dental health and any medication taken regularly, in case it prevents testing of some products. This may be performed confidentially by a health professional, e.g. a company doctor.
- At regular intervals, e.g. annually, check whether the information is correct and note any changes.
- On an ongoing basis, assessors should advise of any changes to their health that have occurred, e.g. new medication.
- During a study, note any unusual health-related occurrences, even if they do not appear to relate the study.
- A mechanism must be put in place to report, record, handle (e.g. 24-hour helpline, visit to healthcare professional) and monitor any adverse events suffered by assessors as a result of testing. An adverse event occurs when the assessor feels unwell or requires a form of remedial treatment. A serious adverse event occurs when the subject requires hospitalisation and may even be fatal. Any adverse event that is experienced during the study must be followed up to resolution.

4.1.4.4 Compensation

Compensation should be available to cover trial-related injury and clinical care, e.g. 'no fault' compensation. Researchers need to have insurance and be indemnified by an employer/sponsor for such claims; however, in the event of negligence, insurance may become invalid. Most sponsors require that they see the insurance. Care must be taken to ensure procedures are in place to keep the insurance valid, e.g. it may be necessary that trials approved by an ethics committee must be signed off by the director of laboratory prior to undertaking the trial.

4.1.5 Safety of test samples

Sample preparation should be carried out using good laboratory and/or hygiene practices (see Section 4.2). When testing is to be carried out by an agency, it is normally the client's responsibility to ensure the samples are safe for testing and this will be laid out in the contract.

4.1.5.1 Sample ingredients

The sensory professional must know the ingredients in the products being tested in order to perform a safety assessment, disclose information to assessors regarding potential allergens (see Section 4.1.5.4) and

Requirements for sensory testing 35

provide information to medical staff in the event of an adverse reaction. In most cases, commercial products have an ingredient list on the packaging and this should be recorded. Ingredient lists and/or formulations for development, and experimental samples, should also be recorded by the sensory study leader.

Caution should be exercised when unknown ingredients are included in testing, e.g. in taint testing. In-mouth and on-skin assessment should not be carried out until the taint has been identified and cleared as safe using a toxicological assessment.

4.1.5.2 Toxicological safety

It is necessary to ensure products and their ingredients are toxicologically safe and they are produced and stored in a manner that avoids toxicological contamination. Generally, products for testing that conform to the legal requirements of the country in which they are being tested can be considered as safe. For ingredients, the type, concentration and process by which they are produced, e.g. ingredients of a genetically modified origin, should be assessed.

Special consideration should be given to products from outside the country in which testing is to be carried out, experimental samples, development samples and samples containing a novel ingredient or an ingredient made using a novel process. A procedure is required to check the toxicological safety of such ingredients and their levels of use and to determine whether clearance is required from an ethical committee, e.g. the approach adopted by the UK Advisory Committee on Novel Foods and Processes (ACNFP 2000) in their guidelines on the conduct of taste trials involving novel foods or foods produced by novel processes (www.acnfp.gov.uk).

4.1.5.3 Microbiological safety

All samples, including samples prepared in the laboratory or pilot plant, to be ingested as part of a sensory test must be cleared for microbiological safety. The exception is products commercially available for purchase, but these must be within their durability date, e.g. 'use-by' or 'best before end date', and be stored appropriately. Microbiological safety can be demonstrated by:
- testing the product to be consumed for microbiological safety;
- approving the product preparation process, e.g. pasteurisation and sterilisation;
- comparing products with established safety standards considering processing, ingredients, packaging, storage, pH, etc.

It is advisable to consult a microbiological expert to determine the most appropriate method.

In all cases, sensory testing should not be carried out unless authorising documentation is physically in the possession of the sensory group. This not only ensures safe samples, but also affords a degree of legal protection to sensory staff and the company should any ill effects occur as a result of undertaking the study.

4.1.5.4 Allergens

Assessors must be notified if the sample contains any known allergens. The European Union (EU) has a list of notifiable allergens that currently includes the following: cereals containing gluten (wheat, oats, barley, rye, kamut, spelt and triticale), peanuts, tree nuts, soya beans, sesame seeds, fish, crustacean, milk, egg, sulphites (at levels above 10 mg/kg expressed as SO_2), mustard, celery, lupin, molluscs and most products or derivatives of these foods. It is also a sensible precaution to avoid the use of compounds known to have a severe allergic effect in some individuals, e.g. quinine.

4.1.6 Safety of experimental/assessment procedure

All aspects of the assessment protocol, from sample preparation to data handling, must be considered for safety. Any procedure that is not consistent with normal conditions should be approved by an ethics committee.

4.1.6.1 Amount of product consumed

Consideration must be given to the overall amount of product consumed or used in the short term (one session) and the long term (over the course of the study or lifetime of the panel) and how any adverse effects may be minimised.

If greater than normal amounts will be ingested/used in one sitting, then consideration must be given to whether the recommended daily intake (RDI) will be exceeded and if so, a safety assessment should be carried out.

Examples of excessive consumption/use that might cause adverse effects are listed here. Strategies to minimise effects are given in brackets.
- Products with a high caloric content may cause weight gain in the long term. (Expectoration rather than swallowing.)
- Alcoholic content may cause slowed reactions and may put assessors at risk of being over the legal limit when driving. (The use of a breathalyser and provision of transport from the test site.)

Requirements for sensory testing 37

- Ingredients that have laxative effects, such as certain fruits, oils, intense bulk sweeteners, fat replacers, and so on. (Minimise consumption in one sitting.)
- Skin sensitivity may be caused by excessive use of some personal care products, e.g. daily shaving, and so on. (Prescreening for excessive sensitivity and minimisation of the length and frequency of use.)
- Sniffing a large number of fragrances in an alcoholic base in one sitting may cause dizziness. (Use sniffing strips and minimise the length and frequency of each sniff and number of products sniffed.)

4.1.6.2 Method of assessment
Any unusual assessment protocols should be assessed for safety and approved by an ethics committee. Examples include the following:
- Any procedure that may cause the assessor discomfort or pain.
- Unusual procedures such as:
  - application of a condition/stimulus for longer periods than usual periods, e.g. extremes of temperature, trigeminal stimuli and so on;
  - putting unusual items in mouth, e.g. very hard samples, temperature-sensing devices, cameras and so on.
- Excessive repetition in the short or long term that could cause lasting effects, e.g. chewing and shaving.

4.1.7 Dos and don'ts
✖ Do not ignore health, safety, ethical and legal requirements. Negligence can have serious consequences for assessors and sensory professionals.
✔ Do become familiar with local and global legislation.
✔ Do take the approach of doing the most you can to uphold professional standards, not the minimum you can get away with.
✔ Do consult experts if you have any uncertainties.

For more detailed information see ASTM E2299-03 (2003), ASTM E1879-00 (2004), ISO 20252:2006, and Pope (1993).

4.2 Good working and laboratory practices

4.2.1 Laboratory practices
As in any laboratory, good laboratory practices, safety procedures and quality procedures should be used. These should conform to local legislation.

4.2.2 Safety

Risk assessments should be carried out on all procedures. Standard operating procedures (SOPs) should be documented. Staff should be trained on the SOPs and a record kept of training received. Control of Substances Hazardous to Health (COSHH) Regulations 2002 assessments should be carried out on all nonfood substances kept in the laboratory, including flavour and odour reference chemicals and cleaning substances.

Fire, accident and emergency procedures, which meet local legislation, should be in place and all staff and assessors should be made aware of them. Particular care should be taken when testing outside normal working hours.

4.2.3 Quality

Quality procedures should be followed, such as the ISO 9000 series (ISO 9000:2000, ISO 9001:2000), MR guidelines, and so on. Quality can be demonstrated by having methodology and training courses accredited, e.g. United Kingdom Accreditation Systems (UKAS), IFST PFSG accreditation scheme for training in sensory evaluation (www.ifst.org.uk).

4.2.4 Special considerations for facilities testing food products

4.2.4.1 Safety

Any area where food is prepared and assessed should be a designated food-safe area. This means that no nonfood chemicals should enter the area. Laboratory coats should be dedicated for use in the area. Glass preparation equipment should be avoided to prevent broken glass entering samples, including glass- and mercury-containing thermometers.

4.2.4.2 Hygiene

Good standards of hygiene that meet local legal requirements must be upheld, which include the following:
- All staff preparing and handling food samples for sensory evaluation must be trained in food hygiene (e.g. hold Basic Food Hygiene Certificate issued by the UK Chartered Institute of Environmental Health).
- All staff must follow a personal hygiene policy, which should include the following:
  - No jewellery, except plain wedding bands.
  - Wear suitable attire: food-safe lab coats, hats (covering all hair), beard snoods, footwear, gloves if appropriate, etc.

Requirements for sensory testing 39

- Wash hands on entering the laboratory and when transferring from one product to another, e.g. meat to dairy.
- Cover broken skin, e.g. cuts, with a blue plaster.
- No person should enter that laboratory if he/she has suffered from vomiting or diarrhoea in the last 48 hours. Certain other medical conditions may require clearance before the laboratory can be entered. It is good practice for staff and visitors entering the laboratory to complete and sign a health declaration form following absence due to sickness or overseas travel outside Europe, USA, Australia or New Zealand.
- An appropriate waste disposal system must be in place, which includes separating biologically unsafe waste, e.g. spittoon waste, from other waste.
- Separate areas and equipment must be used for the preparation of meat and dairy products.
- Samples must be appropriately and hygienically transported and stored. Examples include the following:
  - Samples should be transported and stored at the correct temperature and humidity. Records should be kept of fridge and freezer temperatures.
  - Meat products should be stored separately from dairy products.
  - Products should be covered during transportation and storage.

4.2.4.3 Cleaning
The facility should be easy to clean. It should be maintained in a hygienic condition using cleaning agents that are nonodorous and are not harmful should they come in contact with food and be ingested. Bleach should not be used. Cleaners may need special training on appropriate cleaning regimes. A documented cleaning schedule should be in place and microbiological audits, including swabbing, may be used to check the efficacy of cleaning.

4.2.5 Documentation and data handling
The following types of documentation are recommended:
- Record of the study. It is good practice to document all aspects of a study on a single document or file which details methodology, samples, timing, clearance, analysis, etc. It may be signed off with customers, sensory staff and head of sensory laboratory.
- Laboratory books. Detailed laboratory books of all work carried out should be kept. These should be archived, as they may be needed for legal and patent challenges.

40 Sensory evaluation

- SOPs for all protocols, e.g. sample preparation, hygiene and assessor administrative procedures.
- Guidelines for sensory testing methodology.

4.2.5.1 Data and record keeping

Special procedures are required for keeping and using personal data and test data related to assessors. Procedures must comply with the local data protection legislation and be in accordance with good laboratory and ethical practices. Typical measures include the following:

- Records should be kept in locked draws or protected computer files with limited access.
- Assessor identification codes should be kept separate from personal details. The list of names and associated codes is held on a 'need-to-know' basis only.
- Assessors have a right to view personal information held on them.
- Individual assessors should not be named in presentation of results without consent.
- Data from studies should be archived for a reasonable period of time. For certain types of studies, there may be a requirement to keep the data for longer, e.g. UK ACNFP guidelines on taste trials recommend that records are kept for 30 years.
- Data and personal information must be destroyed in a responsible and appropriate manner.

4.2.6 Intellectual property

Before running a test with nonemployees, and particularly consumers, it is important to consider implications for protection of intellectual property (IP). Using a new product, ingredient or technology in the public domain may invalidate subsequent patent claims, even if consumers have signed confidentiality agreements. Care should be exercised when publishing results prior to being granted a patent, as this can establish 'prior art', which can be desirable or undesirable depending on circumstances.

Sensory testing methods cannot normally be patented, as they are classified as ideas not inventions. Some protection can be afforded by copywriting questionnaires, written protocols, reports and presentations, and registering trademarks for names of unique methodologies. Sensations cannot normally be patented. Patents are based on the compound, ingredient or technology causing the sensations rather than the sensation itself. There is a move towards trademarking signature fragrances and flavours to prevent copies and counterfeiting.

Requirements for sensory testing

Sensory data may be used in patent support. It is good practice to run a specific test to collect data for the patent, rather than try to use data from an existing study, as it must be possible to replicate any test protocol used in a patent so that comparable data are produced. Full details of the test procedure, such that the test can be exactly replicated, may need to be included in the patent or made available to any parties challenging the patent. This may mean fully disclosing an existing proprietary method. It may be preferable to use a simplified version of an existing method, or a methodology tailored specifically for the purpose of patent support, to avoid disclosure.

4.2.7 Dos and don'ts
✔ Do work in an organised, clean and well-documented fashion.
✔ Do adhere to strict hygiene procedures when testing food. Negligence can have serious consequences for assessors and sensory professionals.
✔ Do consult specialists in health, safety, hygiene and IP protection.

For more detailed information see ISO 9000.

4.3 Resources needed for sensory testing

It is essential to determine the type and amount of testing anticipated in the short and long term to assess the resources necessary, including finance, staff, facilities and equipment.

Costs will include initial set-up costs and cost of running testing on an on-going basis. Ideally, it is best to determine needs and then determine the budget necessary to meet those needs. Often, the budget is preset and the scope of the programme and resources must be tailored accordingly. If the cost of testing is being charged to clients, e.g. as an agency on a profit basis or to other departments to cover costs, careful consideration needs to be given to ensure all costs have been included in the charge.

4.3.1 Sensory staff
A sensory team typically includes the following roles: manager, sensory analyst, panel leader, technician, consumer researcher, statistician and assessors. Staff members may fulfil several roles (except the latter role). All sensory staff must have the appropriate knowledge and expertise to carry out their role(s) and should receive appropriate training that may cover the following:
- How to carry out sensory testing, e.g. methods, practices, procedures, data analyses, reporting and so on.

- How to work with sensory and consumer assessors, e.g. motivation, professionalism, confidentiality, safety, ethics, adverse event procedure and so on.
- How to work in a sensory laboratory, e.g. quality (SOPs), safety (fire, COSHH, etc.) and hygiene.
- How to work with clients and project teams, e.g. presentation skills, professionalism and project management.

All too often sensory programmes fail to progress in sophistication because the sensory role is seen as a nonspecialised, nontechnical role that anyone can take on. The sensory role is given to a nonspecialist, who moves on to a different function after 1–2 years, before another nonspecialist is assigned to the role. In reality the opposite is true. The sensory role is highly specialised and technical requiring knowledge of many disciplines including psychophysics, psychology, experimental design, statistics, food technology, physical chemistry, etc. In addition the sensory professional needs to have the ability to work well with clients often from diverse backgrounds and the organisational skills to manage projects to time and cost. It takes many years to become a knowledgeable, well-rounded sensory professional. In order to set up a sustainable programme, in which knowledge is retained and the department grows, it is important to hire and retain technical personnel who want a career in sensory evaluation. There are many excellent university-based degree courses that provide graduates in sensory evaluation. It is also possible to train sensory professionals both through on-the-job training and through methodology-based courses, postgraduate diplomas, etc.

Staff terms and conditions of employment should follow local employment legislation. The human resources department should be consulted on these issues. Legal obligations regarding the terms and conditions of part-time staff employed as sensory assessors should be met, including having a contract specifying remuneration, hours of work, holidays, sick pay, other benefits, etc. Current European law specifies that temporary and casual staff should receive similar rights and benefits as full-time staff. Sensory assessors may also be contracted via an employment agency. This option is more costly, but saves on administration and headcount. It is good practice to have a handbook or charter for sensory assessors which lays out ways of working in the sensory laboratory and the roles and responsibilities of the assessors and sensory staff.

There may be special health and safety considerations when working in a sensory laboratory. For example, if saliva is to be handled, hepatitis and tuberculosis vaccinations are a necessary precaution.

Requirements for sensory testing 43

4.3.2 Testing facilities

4.3.2.1 General considerations

The scope of the sensory testing programme will determine the type and size of facilities needed, e.g. whether the facility will run sensory analysis and/or consumer testing as central location testing (CLT). Whilst it is true that a state-of-the-art, computerised sensory laboratory provides a firm basis for sensory testing, it is not imperative. The most basic of laboratories, where experimental conditions are carefully controlled and experiments are carefully designed, can produce robust and accurate data.

Most facilities (see Figure 4.1 for an example) have several functional areas including a sample preparation area, a serving area, an assessment area with booths, a discussion/training area, staff offices and a storage area. Facilities may also include a reception area, a waiting area, a focus group room and an in-use testing area(s). The overall design should give a flow for movement of both assessors and samples which is both sensible and

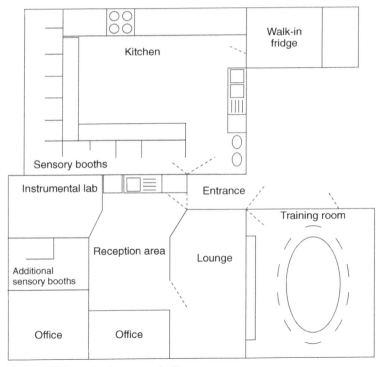

Figure 4.1 Example of a sensory facility.

avoids bias through assessors being able to view samples. Food samples need a hygienic flow to avoid cross-contamination between freshly prepared samples and waste. Ideally, the sample preparation area should be a separate room from the assessment area to avoid bias due to strong odours, noise and overheard conversation related to testing. The facilities should be designed to accommodate disabled assessors.

If new facilities are being built, it is essential to liaise closely with the architect and design team to ensure they fully understand the requirements, to thoroughly check the plans when they are drawn up and to check at regular intervals that plans are being followed. Sensory facilities are unusual in many respects, particularly their bland nature and strictness of specification, and it is not unknown for design teams to make undiscussed changes such as introduce unwanted colour, or downgrade air-handling or lighting specifications.

4.3.2.2 Location

The facilities should be in a location that is odour free and quiet, e.g. avoiding sites that are adjacent to odorous and noisy manufacturing areas, busy roads and so on.

If external assessors are to be used, it is desirable for the facility to be near an area with medium to high population density to ensure an adequate supply of assessors. Ideally, the situation of the facility within the building should be near a building entrance for easy access and near adequate toilet facilities, bearing in mind that an entire panel of assessors may need to use them within a short break time. There should be adequate parking available at the times when assessors arrive for testing sessions.

4.3.2.3 Materials

In order to minimise bias, the facilities need to be as neutral as possible. Carefully consider the use of decorative features, e.g. the colour and content of pictures and patterns may cause distraction and bias, potted plants may generate odour and branded display items may introduce bias. Use nonodorous materials, e.g. surfaces, paints and neutral colours (e.g. pale grey), throughout the facility.

To prevent contamination and aid good hygiene, nonporous, durable materials, such as laminate, plastics and stainless steel, should be used where possible, e.g. work surfaces, chairs, tables and flooring. The used of wood and fabric, which may trap odours, spilt material and dirt should be avoided.

Requirements for sensory testing 45

4.3.2.4 Air handling

Appropriate air handling is particularly important to ensure temperature control and minimise odour build-up. Air handling should be sufficient to counter internal sources of heat, such as computers and lighting, and deal with external variations in temperature, such as extreme heat in summer. It should be able to rapidly clear strong odours such as those generated by cooking or fragrances. The external source of air should be taken into account to ensure it is not being drawn from an odorous source. An odour scrubber or filter may be necessary to clean both incoming and outgoing air. Humidity control may also be necessary. Room temperature and humidity should reflect the conditions under which products are used.

4.3.2.5 Lighting

All testing areas need appropriate and adequate lighting. Simulated daylight is recommended at a standard correlated colour temperature (6500 K) and between 755 and 1070 lux. Care should be taken to ensure even lighting that avoids shadowing in the test area. Coloured lights may be necessary to disguise appearance of samples. This can be achieved by using coloured light bulbs (with no white light component) or filters and, ideally, these should be easily interchangeable with normal lighting. A check should be performed to ensure the coloured light does indeed disguise differences in appearance between samples. For tests to be carried out safely, the intensity of light needs to be at least 300 lux. Light contamination from computer monitors and open serving hatches should be minimised.

4.3.2.6 Functional areas

Reception/waiting area(s)

This is where assessors register their arrival. It often doubles as a waiting area before and between tests. Care should be taken to ensure assessors leaving a test do not influence/bias assessors arriving for a test. Ideally, there should be minimal contact between the two.

Sample preparation area

This is the area where samples are prepared for assessment. Preparation typically includes making many small, identical portions of samples in/on cups, bowls, pots, plates, etc., and laying them out in serving order on trays. It is important that there is adequate working surface space to do this. Adequate space should also be allowed for preparation equipment, e.g. cookers and storage (e.g. cupboards and fridges).

For laboratories testing food, the area must be designed according to strict hygiene design, including hygienic materials (e.g. tiling) and construction (e.g. sealing of cracks and gaps). Ideally, there should be an area at the entrance for storing protective clothing. Lockers may also be useful for storage of personal items that cannot be brought into the lab, such as jewellery. Hand-washing facilities must be present at the entrance.

For laboratories testing both food and nonfood products, and those testing meat and dairy food samples, separate areas must be allocated for sample preparation to avoid cross-contamination.

Sample serving area

This is a space adjacent to the booth area from which samples can be served to the booths. Depending on how serving is to be accomplished, it may need to be wide enough to accommodate trolleys and have enough work space to lay out samples and equipment necessary to keep samples at a certain temperature. It is more convenient to have the height of the serving counter at the same height as the booth counters. Ideally, the serving area should not be visible from the booth area to minimise bias through knowledge of the test and samples. It may be necessary to keep lighting at a lower intensity than the booth area and use the same coloured lighting used in the booths.

Assessment area (booths)

This is the area where assessors work individually in booths to assess samples. Ideally, the booth area should not be directly adjacent to the samples' preparation area to avoid contamination from strong odours. It should have a small positive air pressure relative to the rest of the facility so that no odours are drawn in. If it has windows, these should be covered to avoid external light biasing test results.

Careful consideration needs to be given to the size and design of the booths. These will depend on the type of testing to be performed. It may be desirable to have two booth areas or a large booth area that can be divided flexibly so that more than one panel/project can be run concurrently. Booths may range from partitioned work spaces to small rooms. The size of the booth will depend on the equipment needed during testing, which may include computer hard drives, screens, keyboard and mouse, water, palate cleansers, tissues, trays, plates, etc. Booths should have sufficient electrical outlets and may also include a sink with running water, which should be kept odour free and hygienically clean. The height and design of the booth counters and chairs or stools need to be considered to ensure assessors are comfortable. An assessment should be carried out to ensure safe use of video display units and computer input devices. The booth area should be pale grey or off-white with controlled temperature, lighting and nonodorous air.

Requirements for sensory testing 47

The method of serving samples should also be considered. This can be via a hatch or sliding door, both of which may allow assessors to see into the serving area, or via a pivoting 'bread box' lid or rotating area with central partition, which may restrict sample size.

It is convenient to have a system to allow assessors to communicate with the panel leader to signal for the next sample or completion of the session. This can be achieved using signal lights operated by a two-way switch on both the assessor side and serving side so that the light can be switched off.

Specialised assessment area(s)

For some products, it may be more appropriate to have specialised areas with suitable equipment. Some examples include laundry products that may require washing machines, tumble driers and ironing equipment in a closed area that will capture any odours generated; personal care products that may require individual bathrooms containing sinks, mirrors, showers, etc.; air fresheners that may require a small room; kitchens with hobs, ovens and food-preparation equipment; cleaning products that may require toilets, work surfaces, floors, etc., in a closed area that will capture any odours generated. Assessments may also be carried out under controlled conditions in specialised work environments, e.g. hair salons.

Qualitative discussion groups, e.g. focus groups, are typically run in a room with two-way mirrors, microphones and video-recording facilities. They may also require flip charts, white boards, overhead projectors, video/DVD players, etc. Discussion rooms may also be used to run qualitative groups.

Temporary assessment area(s)

It may be necessary to make sensory assessments outside a central facility, where a sensory laboratory with fixed booths is not available. It is important to ensure the testing room has good air flow and temperature control and is as bland as possible. Portable booths may be used. These often comprise small individual partition walls or three-partition walls that are hinged or slot together. Specialised portable lighting and portable computers can be used. Some agencies offer mobile testing facilities housed in temporary buildings or trailers.

Testing in the home (home use testing)

Testing may be carried out in consumers' homes. This has the advantage that products are assessed under natural conditions of use, but the disadvantage that there is little control over the environment, product and assessment procedure. Home use testing (HUT) presents more of a logistical challenge in that samples and questionnaires need to be delivered to assessors and the noncompletion rate is often higher than for CLT. HUTs are usually more costly and take longer. Classical HUT is carried out with consumers, but it is becoming increasingly common for trained sensory assessors to make assessments at home, particularly when the product is used in circumstances that are difficult to replicate in the laboratory, or when the product needs to be assessed over a period of time, e.g. when testing fragranced bath products.

Discussion room

This is the room where assessors convene as a group to receive training and engage in discussions. It should have a table large enough to accommodate seating, working areas for the assessors and the panel leader, and allow samples, water, palate cleansers, ballots, etc., to be set out. The room should be large enough to allow easy delivery of samples to everyone at the table. A flip chart, white board, overhead projector and computer with network access may be useful. The area should be pale grey or off-white with controlled temperature, lighting and nonodorous air.

Storage areas

It is important to ensure adequate storage space for samples and equipment. Specialised storage such as fridges, chillers, freezers or incubators may be required. Consideration should be given to maintaining appropriate storage conditions, e.g. hygienic conditions, avoiding (cross-) contamination of samples. For storage trials, products may need to be stored under certain conditions, e.g. high temperature, high humidity and daylight.

4.3.3 Equipment

In addition to the equipment already mentioned, specialised sample preparation equipment may be required, e.g. ovens, hobs, microwaves, mixers, dishwashers, measuring equipment such as volumetrics, balances, pipettes, etc.

4.3.4 Dos and don'ts

- ✔ Do plan carefully for immediate and future needs.
- ✔ Do carefully select staff for their role, taking into account their skills, capabilities, experience and long-term motivation.
- ✔ Do hire a sensory professional first and then set up the facilities, rather than vice-versa, so the sensory professional has input into the design.
- ✔ Do visit other sensory facilities to investigate different layouts.

✔ Do allow adequate space for sample preparation and storage.
✔ Do ensure that all resources are in place before starting the test.

For more detailed information see ASTM, Committee E-18 (1986), ISO 13300-1:2006, ISO 13300-2:2006 and ISO 8589:1988.

4.4 Samples

Typically, the objective of a sensory test is to identify the effects of different product treatments. Consequently, it is important that the methods of sample preparation and presentation introduce no additional source of variation into the experiment. However, assessor's safety is paramount and so the general purpose of the test, any use of novel production processes and the ingredients used should be made known to the assessor before testing. Further information on disclosing information to assessors can be found in Section 4.1.

4.4.1 Sample preparation

4.4.1.1 Materials

The materials used for the preparation of samples should be of known origin and storage history. Where several batches of a particular ingredient are required, the batches should be combined prior to use. Where this is not possible, batches should be randomly allocated to products.

4.4.1.2 Equipment and utensils

Care should be taken in terms of the type of materials from which utensils and equipment are made. Glass, glazed china or stainless steel are the most appropriate as they are inert. However, some laboratories may restrict the use of glass due to health and safety issues. Wood should be avoided as it can be unhygienic and can absorb food materials which may be transferred to other samples. Plastic is also capable of absorbing volatile compounds which could then be transferred to later samples. Many standard plastic containers, bags and wraps are not suitable and so should be tested before use. The use of equipment and utensils should be standardised across all samples. Where this is not possible, e.g. different brands of food mixer, samples should be randomly allocated to the equipment.

4.4.1.3 Method

A clear protocol for the controlled preparation of the samples should be documented. All samples should be prepared in exactly the same manner. Procedures should be standardised through accurate use of balances,

50 Sensory evaluation

volumetrics, stopwatches, agitation rates, temperature probes or loggers, etc., where appropriate. Similarly, the positioning of samples in ovens, refrigerators, incubators, etc., should be standardised. If samples are to be stored prior to serving, then the impact of storage on the samples should be tested to ensure that sensory (and microbial) quality is not affected. A preparation method should be chosen which is least likely to mask differences between product treatments. For example, it may be necessary to puree samples to remove textural differences if odour perception is the key objective. However, for consumer testing, a method which closely resembles that used in the home would be the most appropriate.

4.4.2 Sample presentation

4.4.2.1 Sample size and temperature

Careful consideration should be given to sample size, although this may be constrained by the amount of product available. Prior testing with the panel can determine sample size where appropriate. For samples which consist of several components, e.g. casserole, the sample size should be sufficient to enable all components to be included. Clear instructions should also be given to the assessors concerning the sample size that they are to consume.

Serving temperature will be governed by the product and the test objectives. It should be consistent across samples and within a narrow range specified at the beginning of the investigation. The appropriate choice of container can also help maintain sample temperature, e.g. polystyrene cups, provided it does not affect the sensory properties of the product. Water or sand baths, hotplates, vacuum flasks, ice, etc., can be employed to hold samples at the correct temperature provided that the holding time does not affect the sensory properties of the product. Some recommendations are given in the literature for sample temperature. Generally hot foods should be served between 60°C and 66°C, hot tea and coffee between 66°C and 71°C, cold beverages between 5°C and 9°C and frozen desserts between −18°C and −10°C. Many foods can be served at ambient temperature, e.g. snacks, jams and preserves, and cereals.

4.4.2.2 Vessel

The vessel used for sample presentation will depend, to some extent, on sample size. Whichever vessel is chosen, it should impart no additional sensory characteristic to the sample. Glass is preferable, but not always feasible as it is more expensive, requires washing for reuse and can be a hazard. Plastic and cardboard tend to be more popular but should be tested prior to use to ensure they do not affect the sensory properties of the product, e.g. avoid waxed board. Identical vessels should be used for all products and the nature of the vessel should not bias the assessor. For example, using ornate wine glasses may bias the perceived quality of a beverage. Clear or plain white vessels are recommended unless the vessel is also being used to mask unwanted differences between products, e.g. dark opaque glasses are often used to mask differences in appearance between wines.

4.4.2.3 Carrier

Some products, e.g. fat spreads, breakfast cereals, seasonings and flavourings, are not readily evaluated on their own and require an additional product to be presented as a carrier. This is particularly the case in consumer tests in which presenting a product as it is normally consumed is especially important. Careful consideration should be given to the choice of carrier as it can provide an additional source of variation in the data. Furthermore, interactions between the product and the carrier (physicochemical and/or perceptual) may result in the evaluation of sensory sensations not uniquely related to the test product. For example, the presence of bread may influence the transport of taste and aroma stimuli to the receptors, the viscosity of a sauce is known to influence flavour perception. The sensory analyst should carry out preliminary investigations to ensure that the carrier does not distract assessor attention from the test product and does not influence its sensory character, either by adding to or by masking its sensory attributes. Some typical carriers include unsalted crackers, white bread, rice, pasta, milk and bland white sauces. The carrier, however, should always be congruent with the test product, e.g. bread is useful for spreads but not sauces intended for use with fried noodles.

Some nonfood products may also be best assessed on carriers, particularly those containing a fragrance. Carriers include blotters (fragrances and aerosols/sprays), skin (perfumes, creams, washing products, shaving products, etc.), hair and hair swatches (hair care products), towels and linen (laundry products), crockery (dishwashing products), etc.

4.4.2.4 Coding

Coding samples removes a source of expectation error from the experiment. Codes used for samples should be carefully recorded and random three-digit codes are preferred. They can be randomly generated by computers and calculators or obtained using random number tables, although certain codes should be avoided (see Section 2.2.1.1). Different codes should be used for replicates to prevent assessors remembering samples. All codes should be written in a consistent format and positioned similarly using the same odour-free pen or printed by computer on labels subsequently positioned on the presentation vessel.

4.4.2.5 Palate cleanser

To avoid carry-over effects and adaptation to sensory stimuli, the use of a palate cleanser between samples is required. Bottled mineral water at room temperature is successful in cleansing the palate for a wide range of products (the flavour of tap water tends to be too inconsistent). However, some products can be more problematic. Greasy foods tend to need something more astringent, e.g. pieces of apple can be useful between samples of chocolate. Spicy foods tend to need a palate cleanser containing fat, so milk is a popular choice. For astringent products like tea, melon is successful. The sensory analyst should determine an appropriate palate cleanser prior to the investigation with input from the panel where appropriate. The time given between samples can also be important for products with lingering effects. Although it can be hard to police the use of palate cleansers, it is important that assessors are instructed to be consistent in their own approach. Consistency across the panel may be more difficult to attain.

4.4.2.6 Number and order of samples

Balancing the order of sample presentation removes several sources of error (see Section 2.2.1). The number of samples to present will depend very much on the nature of the samples, the size of the sample, the objective and type of test, and the time available. For visual assessments, tens of samples can be assessed in any one sitting, whereas for some spicy samples only one product per session may be possible.

4.4.3 Reference samples

Reference samples can be used to exemplify an attribute or a particular intensity point on a scale. The purpose of the reference should be made clear to the assessor. The same controlled procedures described earlier are required for the production of any reference samples, particularly when different batches need to be made. If the reference is at room temperature, it may be left in the booth for the duration of an experiment. Some references, however, will need to be served at specific temperatures and hence renewed each time they are used. The use of a reference sample should be standardised across assessors so as not to add a further source of variation to the experiment. The reference may also be presented as a coded sample to evaluate the consistency of the assessors.

Requirements for sensory testing 53

4.4.4 Sample assessment procedure

The protocol for evaluating samples should be determined by the test objectives. The protocol must be clearly indicated to the assessors in a briefing session and/or in the instructions presented with each sample. If assessors are not consistent in the manner of their assessments, it will add an additional source of variation to the data.

The decision of whether assessors should be asked to expectorate (spit out) the sample needs careful consideration. Expectorating samples can enable more samples to be evaluated, and may be particularly favourable where the ingestion of alcohol or high amounts of fat is concerned. The act of swallowing, however, can be very important to the sensory properties of many products. Swallowing plays a key role in the delivery of aroma volatiles to the nasal cavity, and hence flavour perception. Furthermore, many sensory receptor cells are also present in the back of the oral cavity and on the throat. Indeed, important attributes for many products are sensed here, e.g. the 'burn' associated with carbonated beverages. In consumer tests, where consumption should match normal eating conditions, swallowing the sample becomes particularly important.

Clearly defining how the sample should be consumed or used is of utmost importance in any sensory test. For some sensory methods, defining how the sample should be assessed is inherent in the method itself, e.g. descriptive profiling, but for others it is up to the investigator to ensure assessors are informed of the assessment protocol. For food products, consideration should be given, for example to how the product should be placed in the mouth, whether specific teeth are used for the first bite and whether the number of chews should be specified for solid foods. For nonfood products, clear instructions are just as important, e.g. stating how long to brush with toothpaste, to what size area skin cream should be applied and how to apply it, how to sniff to assess perfumes and air fresheners.

The timing of when to make the assessment of a particular attribute should also be clear, e.g. when assessing initial flavour intensity a judgement needs to be made on first contact with the sample, whereas overall flavour intensity may be judged once the product is consumed.

4.4.5 Dos and don'ts

✔ Do check that the samples received are what you expected. If the samples have been specially made, check that the sensory and physicochemical properties are as expected. If the samples have been purchased from the market, ensure they are the correct products from a single batch.

54 Sensory evaluation

- ✓ Do seek to understand the variability in your samples prior to the study. Large variability within the product will make it difficult or even impossible to assess differences between products.
- ✓ Do fully evaluate the samples, including carrying out a tasting with the sensory team, prior to the study. This will help them better plan the study, identify extremes, verify the samples and understand the results.
- ✓ Do carefully consider how to handle extreme samples, as they may skew the data, obscuring the main effects. Consider making extremes a block at the end of the design, or omitting them from the study.
- ✓ Do be cautious about choosing to leave out the control when the number of samples that can be tested is limited. It may be a false economy.

4.5 Assessors

The individuals who take part in product assessments are the most important asset in sensory evaluation. It is essential that the most appropriate individuals are recruited, screened and selected to take part in sensory tests; they should be given the necessary training and tools to allow them to complete the test effectively; they should receive positive and constructive feedback on their performance and, above all, they must be treated with consideration and respect.

4.5.1 Recruitment

The method of recruitment will depend on the number and type of assessors needed. When recruiting a trained panel for profiling, the approach may be different to that used for recruiting consumers onto a database or for participation in an affective test. Assessors may be prerecruited in advance of the test or recruited immediately prior to the test as an integral part of the test protocol.

4.5.1.1 Advertisement

Adverts can be placed in local papers, on notice boards in public places, handed out in person, posted by hand or mailed out. This is an efficient way of reaching a large target audience and is useful for all recruitment purposes.

The advert must be clear and unambiguous, conform to local employment law, and give details of what is expected from individuals and what action they must take if they are interested in participating in sensory tests. Be aware of the consequences of the advert, e.g. telephone contact details can result in switchboards/telephone lines being jammed. Staff must be available to deal with potentially large volumes of respondents and the associated administration.

4.5.1.2 Direct recruitment

It is possible to recruit people directly, either face-to-face or by telephone. This type of approach is particularly effective when recruiting consumers for participation in affective tests. The recruiter should be a registered member of a market research organisation, e.g. ISOMAR, and be able to provide necessary credentials.

4.5.1.3 Word of mouth/recommendation

Recommendation from existing participants can be a very effective method of adding new 'consumers' to a database or introducing new members to a trained panel.

4.5.2 Internal vs. external assessors/panels

Irrespective of the ideal, a decision needs to be taken on whether to use:
- internal assessors/panel of existing employees;
- externally recruited assessors/panel who are employees dedicated to the role;
- externally recruited assessors/panel from an employment or sensory testing agency.

Constraints, such as available time and money, will impact on this decision. Whichever choice is made, it is vital that any limitations or consequences be noted. This is most important when interpreting data and drawing conclusions that may have a large financial impact. Table 4.1 highlights some of the advantages and disadvantages associated with each option.

Internal panels can be effective for simple discrimination tests. Problems may arise when they are needed for lengthy descriptive studies, as individual line managers may not be willing to release staff for long periods of time and motivation often declines as employees become less willing to give panel work priority over their own work.

Owing to their product knowledge, bias towards company products and nonrepresentation of the population, employees should not be used for affective tests. In certain circumstances, often related to budget, some companies do use internal employees for simple hedonic studies, but such data should be treated with caution and used only as a guide to consumer trends.

Table 4.1 Summary of advantages and disadvantages associated with internal and external panels

Panel	Advantages	Disadvantages
Internal	• Perceived as less expensive • Perceived as readily available • Flexible • Requires little notice • Confidentiality retained in company	• May actually be more expensive • Distracted by full-time job – may not make sensory testing a priority • May have too much product/project knowledge • Biased/conflicted • Less time available for training • Not invested in test • Not available together • Often, only available for short tests
External	• Readily available • All available together • Properly trained • Invested in the task	• Perceived as more expensive • Disruptive group dynamics • Sociable aspect can interfere with work • Require constant supply of work for efficient operation

4.5.3 Screening and selection

All potential participants, irrespective of how they are recruited, must be screened prior to final selection for the panel. This screening process determines that individuals satisfy some minimum criteria for selection. These criteria are very dependent on the type of test and may even be as simple as personal availability. Remember that the more complex the selection criteria, the more difficult, time-consuming and expensive it will be to find appropriate participants. Upon selection, a minimum amount of information about assessors must be collected and held according to local data protection legislation, which, in some countries, must include name, address and social security number for tax purposes.

4.5.3.1 Naive assessors
Untrained sensory panel
This type of panel participates in very simple sensory tests, e.g. a paired comparison test, and, therefore, the required level of screening is minimal. A naive panel often includes individuals who are willing to take part in the assessment and are available to do so. Screening for sensory acuity depends on the test objective. Participants often undertake various screening tests

upon selection for the panel, so that their suitability for different types of tests is known in advance.

Consumers
Consumers may be selected for a one-off, affective test, a consumer panel used to provide affective judgements on a frequent basis, or for a database from which consumers are drawn to participate in affective tests on a less frequent basis. The screening for selection typically involves a number of simple questions, e.g. personal details, demographics, purchase behaviour, current product use and previous participation in consumer tests. The type and number of questions can vary depending on the objective. This type of screening often occurs at the same time as recruitment or, in the case of existing consumer databases, may precede telephone recruitment.

4.5.3.2 Trained panel
Individuals on a trained panel must exhibit suitable personality and attitude traits, be in good health and satisfy minimum criteria for sensory acuity (ability).

Personality
Generally, participants should:
- be able to work as a team;
- have cosmopolitan preferences;
- be positive but not overbearing;
- be a good listener and communicator;
- be committed;
- be flexible.

Health
Assessors must be in good general health; any physiological or health restrictions must be documented, e.g. allergies, false teeth, migraines, as these may affect their participation in certain tests.

Sensory acuity
Assessors should have at least normal sensory acuity with regard to:
- detecting stimuli;
- discriminating between stimuli;
- recognising and describing stimuli.

In some circumstances, e.g. taint assessment, it may be desirable to select assessors with high sensitivity to certain stimuli.

ISO 8586 parts 1 (1993) and 2 (1994): 'General guidance for the selection, training and monitoring of assessors' details specific tests used to determine individuals' ability to detect, discriminate and describe. In some instances, it is best to perform a general screening of abilities for appearance, texture, taste and aroma. However, when the panel is to be used for one specific purpose, e.g. texture assessment, the screening activities can be tailored to suit these needs.

Detecting stimuli

Identifying and scoring the intensity of specified attributes is an integral part of many sensory tests and, therefore, it is important that assessors can detect visual, odour, tactile and taste stimuli at typical concentrations/intensities.

Common methods used to assess detection ability are as follows:
- *Threshold tests*: Increasing/decreasing concentrations of a stimulus are presented. These are commonly used for basic tastes.
- *Colour blindness tests*: Ishihara or Munsell-Farnsworth tests can be used to determine a large number of visual impairments.

Discriminating between stimuli

Discriminating between samples that may vary in intensity of certain attributes is a fundamental requirement for many sensory tests. It is important that assessors be sufficiently sensitive to varying concentrations/intensities of these attributes.

Common methods used to assess discrimination ability are discrimination tests (see Section 5.2): paired comparison, triangle and ranking test. Typically, the samples used in screening tests are representative of those the panel will encounter in the future, e.g. if they were to assess carbonated beverages, screening samples could include different levels of sugar, acid or aroma compounds. It is also important to ensure that assessors can pick out stimuli in a mixture and in the product to be tested. Careful consideration needs to be given to the differences between samples. If they are too great, the test will be too easy whereas if they are too similar, the test will be too difficult and stressful.

Recognising and describing stimuli

Although recognition of sensory properties will form part of training, it is important that assessors have a basic ability to recognise and name stimuli. Typical tests include presenting a series of basic tastes and/or odours and asking assessors to name the sensations. Table 4.2 lists typical examples and concentrations used for taste and aroma recognition tests.

Requirements for sensory testing 59

Table 4.2 Examples of samples and concentrations used for determining recognition levels for taste and aroma compounds

Property	Material	Concentration (g/L)
Taste[a]		
Sour	Citric acid	0.43
Bitter	Caffeine	0.195
Salt	Sodium chloride	1.19
Sweet	Sucrose	5.76
Umami	Monosodium glutamate	0.595
Metallic	Iron(II) sulfate heptahydrate	0.00475
Orthonasal aroma[b]		
Fresh lemon	Citral	0.01
Rose	Geraniol	0.01
Grass, green	Cis-3-hexen-1-ol	0.05
Almond	Benzaldehyde	0.05
Strawberry, banana	Ethyl butanoate	0.005
Rancid, cheesy	Butyric acid	0.01
Cloves	Eugenol	0.005
Mashed potato, grilled meat	Methional	0.01

[a]From ISO 3972:1991.
[b]From ISO 5496:1992.

Texture recognition is also important; samples selected to assess this tend to be those relevant for the industry in question.

The ability to describe product attributes, and effectively communicate these ideas, is a necessary skill for assessors involved in sensory profiling methods. Running a 'mini profile' is a useful way to assess this ability. Individuals are asked to record on paper words or phrases that describe the appearance, aroma, texture, taste and aftertaste of 2–3 products. The products are selected to provide as much variability as possible. Following this they are directed by a facilitator to discuss their perceptions with the rest of the group. This allows judgement of not only the descriptive ability but also important personality traits (see Section 4.5.3.2).

Interpreting results

Interpreting the results, and deciding if an individual is suitable, depends on the type of testing the individual will be required to carry out. For example, assessors may be insensitive to bitter compounds or unable to distinguish some shades of blue and green; however, if they will not be required to assess samples with these attributes, then exclusion may be unnecessary.

Decision criteria may be 100% correct responses for attributes deemed to be very important; this may be reduced to 60–80% correct responses for others. Typically, only 10–30% of participants would be expected to pass screening for a trained sensory food panel. Documenting assessors' strengths and weaknesses provides useful information for future changes in panel use.

4.5.4 Training
The level of training required by each assessor is driven by the test method itself. In some instances, no training beyond the instruction required to complete the test is necessary; whereas, in other cases lengthy training on attributes and scales may take several weeks/sessions to complete.

4.5.4.1 Discrimination tests
Participation in these methods is generally considered to be straightforward and assessors often require familiarisation, rather than in-depth training, although attribute-specific discrimination tests may require training on the test attribute. The simplest techniques, e.g. paired comparison, triangle and duo-trio test, can be performed with naive or trained assessors depending on the test objective. Some texts state that the recommended minimum number of participants is dependent on assessor type.

Assessors should be given clear instructions on how to complete the test. Attention should be drawn to the consequences of not following the protocol, e.g. poor palate cleansing, not assessing samples in the order presented. These instructions should be given to all assessors irrespective of their previous experience.

A 'practice test' can be run for panels that are participating in tests for the first time. This increases understanding and eliminates anxiety.

4.5.4.2 Descriptive tests
General training should be provided for assessors participating in descriptive techniques. This should precede any method-specific training, which is usually intensive and is covered in detail in Section 5.3.

The purpose of general training is twofold. Not only should it enhance detection, discrimination and descriptive skills but also build self-confidence and reduce anxiety.

Training methods are, typically, an extension of those used for screening. As the assessors gain experience, the methods should increase in complexity, e.g. the samples used in triangle, duo-trio or ranking tests can be more difficult to discriminate. If a particular product type is to be assessed, this should be included in training so that attributes become familiar. The mini profile (described earlier) is a useful training tool that provides the assessors opportunity to become familiar and confident with the test protocol.

Requirements for sensory testing 61

In some instances, general training is combined with the method/product-specific training performed for each study. This is most common when each study involves a very different product group and/or test method.

4.5.4.3 Affective tests

Training for affective tests requires only a clear description of the test method. No assessor training is needed when questionnaires are filled in by an interviewer during a face-to-face interview, although guidance may be needed for computer-based questionnaires.

4.5.5 Motivation

Motivating the panel will improve/maintain the quality of assessors' data. Whilst money may be the reason that assessors attend a session, it is motivation and the desire to do a good job that keeps them focused and invested in the task. There are several methods that can be employed to improve motivation.

4.5.5.1 Feedback

Whenever possible, provide feedback about the work the panel have performed. This can happen throughout the project although, to avoid bias, certain information may be restricted until the end. Feedback should be positive and can include information about their performance; comments from the client; summaries of presentations, publications and technical reports; actions taken as a result; future applications and work. Feedback can be given in person, through production of a newsletter, or as an organised event, e.g. meeting or mini symposium. It may also form a part of an annual appraisal.

4.5.5.2 Personal contact

Taking the time to talk to the panel about projects, keeping them up to date with information about the company or, simply, taking the time to get to know them demonstrates a level of respect that is inherently motivating. Making a personal connection with the panel will result in a desire to do a good job for a known person rather than an anonymous company. It also offers the panel a chance to give feedback and make suggestions.

4.5.5.3 Group activities

Organising group activities can be very effective, particularly when all assessors do not work together at the same time. Activities can be related to work, e.g. visiting factories and other sensory panels, or purely social,

62 Sensory evaluation

e.g. theatre, dinner and parties. When the activity is organised and subsidised by the organisation, it better demonstrates the value placed on the panel work and commitment.

4.5.5.4 Remuneration

Assessors are often paid for their participation. This may range from a nominal payment for a one-off test to a salary for regular participation in a trained panel. If assessors are company employees who are participating in sensory tests outside their normal role, recognition for attending sensory tests becomes a particularly important motivating factor. Techniques include additional monetary payment, gift certificates, bonuses for completing a certain number of tests, inclusion in a raffle, sweets/cakes/cookies after the test, free lunch, etc. In all cases, the tax implication of such remuneration needs to be considered.

4.5.6 Good working practices for assessors

Generally, the following good working practices should be observed by all assessors irrespective of training and test protocol.
- Assessors should not smoke for at least 1 hour prior to the start of a food or fragranced product test as this affects their sensitivity to certain attributes and creates lingering odours that distract other assessors.
- Assessors should not wear highly fragranced, personal care products or cosmetics, as this may interfere with the product assessment.
- Assessors should not eat or drink for at least 1 hour prior to the start of a food or fragranced product test.
- Assessors should not talk during a test unless instructed to do so.
- Assessors should observe good personal hygiene, e.g. body odour may distract coworkers.
- Assessors should attend on time.
- Assessors should focus on the test and follow instructions.

4.5.7 Monitoring panel performance

The performance of assessors should be monitored as an integral part of any project. Furthermore, ongoing sensory acuity and capability should be assessed periodically as part of a long-term monitoring programme. The most important criteria for panel performance are accuracy, precision and reliability.

Accuracy: This is the measure of how close assessor data or panel mean data are to the *true* value. A true value is not always easy to identify; where possible, spiked samples and references can be used.

Precision: This is a measure of how reproducible assessor or panel data are, i.e. how close are the replicate scores/judgements or mean data.

Requirements for sensory testing 63

Reliability/validity: This is a measure of how close an individual assessor's score/judgement is compared to the rest of the assessors and panel mean. The way in which these criteria are assessed is dependent on the test method and data type.

For discrimination tests, monitoring performance is a very simple process, i.e. can an assessor correctly discriminate between the samples? are replicate discrimination tests (where used appropriately) reproducible? are individual assessors consistent with one another? For individual projects, where the expectation is that the samples are confusable, these questions are not appropriate. They do apply, however, for an ongoing monitoring programme in which assessors receive the same sample sets every month, 6 months, etc. and are expected to provide consistent results demonstrating discriminative ability.

For descriptive tests, the data and, therefore, the mechanisms by which panel performance can be judged, are more complex (see Section 5.3.2.7).

4.5.8 Dos and don'ts

- ✔ Do ensure screening tests are appropriate and relevant to the study.
- ✘ Do not rely only on model samples for screening, e.g. water-based solutions of basic tastes – use real, relevant product samples also.
- ✔ Do pilot screening tests to check whether they provide an appropriate level of screening, i.e. they are not too easy or too difficult.
- ✔ Do use the correct type of assessors.
- ✘ Do not use a trained sensory panel or quality panel to give consumer liking, pleasantness, preference or acceptability ratings.
- ✘ Do not use consumers to provide objective sensory measures.
- ✔ Do test the appropriate consumer group with emphasis on the target consumer.
- ✘ Do not assume that allowing familiarisation with a product or procedure is the same as providing training.

4.6 Data capture

Various means of data capture are available to the sensory professional, the choice of which should be directed by project objectives and financial considerations.

4.6.1 Paper

Traditional pen and paper offers several advantages, not least that it is relatively inexpensive and not as susceptible to technical glitches as computerised data collection. In addition, the design of a paper response sheet requires little or no training in computer software. Paper response

64 Sensory evaluation

forms are also highly mobile and can be used at numerous locations and with almost all types of consumers without the need for an additional power source or technology. On the downside, designing paper response forms with balanced orders of presentation can be laborious, forms can be lost, errors can occur in transferring the data to computers for statistical analysis and the purchase of additional software for data analysis is almost always required.

4.6.2 Computerised systems

Computerised systems for data capture are now the norm, and several companies produce bespoke software for sensory investigations. There are many advantages to this means of data capture. Although initial training is required, considerable time is saved when designing and setting up sensory tests as the software takes care of much of the experimental design for various sensory methods and automatically saves the data in a format ready for data analysis. Most computerised systems have built-in statistical software which enables rapid analysis and reporting of the data. Computerised systems do, however, require assessors to have basic keyboard and mouse skills. Initial set-up costs can be high, and computers are vulnerable to power failures and technical glitches which can result in data loss or the inconvenience of cancelled panels. As light from a monitor can affect test conditions, their use in the evaluation of products where lighting parameters are important should be carefully considered, e.g. assessment of the appearance of hair colour swatches.

4.6.3 Portable systems and the internet

The availability of laptop PCs and hand-held personal organisers means that sensory data can be captured electronically at central location tests, in the home and even in mobile sensory units. The advent of wireless technology and the internet also means that such data can be quickly downloaded to a remote central database from several locations for statistical analysis.

As the number of homes with internet access has increased so has the capability for large-scale, potentially global, consumer investigations. Several companies now offer the facility to set up web-based consumer surveys with access to thousands of consumers who have already volunteered to participate. Limitations of such studies include that the 'sample' is restricted to those who have internet access and that it may be difficult to verify the identity of the person answering the questionnaire.

The internet can be used for traditional consumer-type questionnaires and, more recently, conjoint analysis type studies. Qualitative work may also be carried out through the use of chat groups or forums. Furthermore, the internet also provides opportunities for trained sensory panel to assess products 'at home' and complete response questionnaires online. As the web, and the electronic technology in general, continues to develop so will the opportunities for sensory and consumer testing.

4.6.4 Qualitative research

Where the objectives of a study are more qualitative, focus groups and ethnographic research techniques are more likely to be utilised to obtain sensory-related data. (Ethnographic research is based on descriptive observations of respondent behaviour by the researcher.) Data capture, in this instance, is likely to take the form of written notes and/or involves the use of tape and video recorders and cameras.

5 Sensory test methods

5.1 Selecting the test

Sensory test methods are designed to answer the following questions: Is there a difference? What is the nature of the difference? Is the difference acceptable?

There are two types of sensory tests: objective and subjective.

Objective tests provide objective data on the sensory properties of products and are carried out by trained assessors. There are two classes of objective tests:
- *Discrimination tests*: Determine whether there are sensory differences between samples.
- *Descriptive tests*: Identify the nature of a sensory difference and/or the magnitude of the difference.

Subjective tests are known as affective or consumer tests. They provide subjective data on acceptability, liking or preference, and are carried out by untrained assessors.

This chapter gives detailed information on discrimination, descriptive and affective tests, including their objective, procedure, experimental design, questionnaire, data analysis, conclusion and an example.

5.2 Discrimination tests

5.2.1 Introduction

Discrimination tests are some of the most common methods employed in sensory science. They are used to determine if a difference (or similarity) exists between two or more samples. Statistical significance testing is used to analyse the data and determine whether or not samples are deemed to be different or similar.

Discrimination tests are rapid techniques and can be performed by both naive and experienced assessors; however, a panel should not be a

combination of both. These are often used when the samples are considered to be 'confusable', i.e. their differences are not obvious but need to be investigated. They are commonly used in the following circumstances:
- Screening and training assessors
- Investigating taints
- Determining sensitivity thresholds
- Quality assurance/quality control, e.g. screening raw materials for consistency
- Investigating the effect of ingredient/process changes, e.g. for cost reduction or supplier change
- Preliminary assessments

There are several International organisation for standardisation (ISO) and American society for testing and materials (ASTM) standard methods for discrimination tests (www.iso.org; ISO 8588:1987; ISO 8587:1988; ISO 4120:2004; ISO 10399:2004; ISO 5495:2005; www.astm.org).

5.2.1.1 Setting objectives for the test
To avoid confusion and disappointment, it is necessary to determine specific objectives for the test. These, along with other considerations of sample, timescale and cost, will affect the choice of test method.

Furthermore, it is important to understand the limitations of a discrimination test and consider these against the test objectives. For example, a standard triangle test (see Section 5.2.2.1) may determine if a significant difference exists between two samples but, used alone, will neither give you information about the degree of difference nor indicate which sample is preferred.

5.2.1.2 Test environment
Discrimination tests are typically carried out in tasting booths or a similar environment that is free from bias.

5.2.1.3 Re-assessing samples during the test
It is important to specify whether or not assessors are permitted to re-assess the samples before making a judgement. This is a matter of choice and will be determined by factors such as the quantity of sample available, the nature of the samples, carry-over effects, number of tests to complete (minimising fatigue) or the purpose of the test.

5.2.1.4 Forced choice vs. no difference
When setting up a discrimination test, you must decide how the assessors are allowed to respond. The 'forced choice' mode dictates that a

68 Sensory evaluation

decision must be made and a sample selected in response to the question, e.g. which sample is the 'sweetest' or the 'odd one out'. The 'no difference' option allows the assessors to report that the samples do not differ with regard to the question asked.

There is some debate concerning which of these options is most appropriate to use. For example, a trained and experienced panel may resent being forced to make a choice when they perceive the samples to be the same. In contrast naive assessors will often select the 'no difference' option rather than risk making a choice, just in case it is wrong, or because they are not motivated to look for a difference.

If a 'no difference' option has been allowed, there are three possible approaches to the data analysis; the third option is rarely used.

1 Ignore the 'no difference' responses. This will reduce the number of assessors and, consequently, reduce the power of the test. The number of 'no difference' responses should be reported.
2 Split the 'no difference' responses proportionally between the products using the assumption that if assessors were forced to make a choice, their results would be randomly split. The number of 'no difference' responses should be reported.
3 Distribute the 'no difference' responses proportionally according to the rest of the data in which a choice had been made. Essentially, this implies that a forced choice approach should have been implemented in the first place.

5.2.2 Overall difference tests

In overall difference tests, assessors can use all available information to make their judgement. In some instances, the tests can be restricted to one modality, e.g. appearance or aroma; however, this will require the disguise of other sample attributes. It should be noted that it is not acceptable to simply instruct the assessor to focus on one modality; different means of disguising the other stimuli are commonly used. For example, coloured lights can disguise the appearance of samples when visual differences would make it easy to determine the 'odd' sample. It may be that the project leader is interested in all aspects of texture, aroma and flavour and does not want appearance to be assessed. It is essential that any form of disguise be thoroughly checked to ensure that it is effective; otherwise final conclusions may be based on incorrect assumptions about the samples.

5.2.2.1 Triangle test

Objective: To determine if a difference exists between two samples.

Sensory test methods 69

Procedure: Assessors are presented with three samples and told that two samples are the same and one is different. They are asked to assess the samples in the order provided and determine which sample is 'the odd one out'. They may also be asked to describe the difference. Appropriate palate cleansers should be used between each sample. Samples are labelled with three-digit codes (blind coded).

Experimental design: There are six possible orders of sample presentation. They are

AAB   BBA
ABA   BAB
BAA   ABB

In some instances, only one half of the design is used, for example if the quantity of one of the samples is limited, or if one of the samples is the standard/reference and, therefore, presented as the duplicate sample. It is good practice to use each possible presentation order an equal number of times with 24–30 assessors, although the absolute number chosen depends on the overall aim and the significance level selected. Larger panels are more discriminating and are commonly used when the differences are very small or when the aim of the test is to determine similarity (see Section 5.2.4).

Questionnaire: See Figure 5.1.

Triangle test

Assessor: Date:

You are provided with three samples, each labelled with a three-digit code. Two samples are the same and one is different. Assess each sample in the order provided, from left to right, and select the 'odd' sample. Record your result below.

Cleanse your palate with cracker and water after each sample. You are not permitted to retaste the samples. Please comment on how the odd sample is different.

Sample	Different sample (please tick)
219	
470	
593	

Comments:

Figure 5.1 Example of a questionnaire for the triangle test.

70 Sensory evaluation

Data analysis: The total number of responses correctly identifying the 'odd' sample is counted. There are two ways of analysing the data.

If analysing the data by hand, the number of correct responses is compared to statistical tables (see Appendix 3). The table states the *minimum* number of correct identifications required (at different levels of significance) before a significant difference can be concluded from the test. The total number of correct responses must exceed the critical minimum value from the table.

Alternatively, software packages calculate the probability of making a type I error (α risk) should it be concluded that a significant difference exists between the samples. In this instance, a probability of less than 0.05 (equivalent to 5% level of significance) is used as a 'cut-off' although common sense should be used when interpreting this data, e.g. would it be sensible to ignore a result of $p = 0.056$ and conclude that no significant difference exists just because this value is not less than 0.05?

Conclusion: From a triangle test, the conclusion is that a significant difference does OR does not exist between the two samples. In either case, the significance level of the test, e.g. $p = 0.05$, must also be stated. In addition a comment may be made about the nature of the difference.

Example: A juice company was considering switching suppliers of apples. The action standard for making the switch was no significant sensory difference at the 5% level between the new supplier's apple juice and the current supplier's juice. The company decided to run a triangle test with an objective to determine if a significant difference existed between two batches of juice made with apples from the two suppliers. The significance level chosen for the test was 5%. Twenty-four untrained assessors participated in the triangle test; numbers were kept to a minimum to save money. Sixteen assessors correctly identified the 'odd' sample.

From the table in Appendix 3, for a panel of 24 assessors, the minimum number of correct responses required at 5% significance level ($p = 0.05$) is 13.

From software packages, the probability of making a type I error with this result is $p = 0.0009$. This is less than $p = 0.05$ (the significance level of the test); in fact it is also less than a 0.01% level of significance.

Conclusion: There is a significant difference between the two batches of apple juice ($p < 0.05$). The action standard has not been met and suppliers will not be switched.

5.2.2.2 Duo-trio test

Objective: To determine if a difference exists between two samples.

Procedure: Assessors are presented with three samples, two blind coded and one labelled as a 'reference'. They are asked to assess the reference sample, followed by the two coded samples (in the order provided) and determine which is the most similar (or different) to the reference. Appropriate palate cleansers should be used after each sample. The duo-trio test is particularly useful for samples that are not homogeneous, as the question asked is, which sample is the 'most similar' (rather than 'identical') or 'most different' to the reference.

Experimental design: There are four possible orders of presentation in which either sample can be used as the reference. These are

        Ref A   AB
        Ref A   BA
        Ref B   AB
        Ref B   BA

There are two possible formats for the duo-trio test.
1 Balanced reference technique in which all four possible orders of presentation are used and the reference can be either sample.
2 Constant reference technique in which only two of the possible orders of presentation are used and the reference is always the same sample. The constant reference technique can be employed for various reasons, e.g. comparing products to a gold standard where the reference is well defined, where one of the samples has a limited quantity or where one sample is particularly well known to the panel.

In its standard format, this test (and other discrimination test methods) requires the assessor to remember differences between samples that are not adjacent in the tasting order. The presentation design can be modified, in which the reference sample is presented in between the two test samples. This format minimises the effect of memory, as the assessor needs to remember only the difference between the test sample and the adjacent reference. Larger panels are statistically more discriminating of smaller differences between samples.

It is good practice to use each possible presentation order an equal number of times with a minimum of 32 assessors, although the absolute number depends on the overall aim and the significance level selected.

Questionnaire: See Figure 5.2.

72 Sensory evaluation

Duo-trio test

Assessor: Date:

You are provided with three samples, one is labelled as the reference (REF) and two are labelled with a three-digit code. Assess the reference sample followed by each coded sample in the order provided, from left to right, and determine which is the most similar to the reference. Record your result below.

Cleanse your palate with cracker and water after each sample. You are not permitted to retaste the samples. Please comment on any differences between the samples that you experienced.

Sample number	Most similar to REF (please tick)
036	
619	

Comments:

Figure 5.2 Example of a questionnaire for the duo-trio test.

Data analysis: The total number of correct responses (correctly identifying the sample that was the same as or different to the reference) is counted. There are two ways of analysing the data.

When analysing the data by hand, the number of correct responses is compared to statistical tables (Appendix 4). The table states the *minimum* number of correct identifications required (at a specified level of significance) before a significant difference can be concluded from the test. The total number of correct responses must exceed the critical minimum value from the table.

Alternatively, software packages calculate the probability of making a type I error should it be concluded that a significant difference exists between the samples.

Conclusion: From a duo-trio test, the conclusion is that a significant difference does OR does not exist between the two samples. In either case, the significance level of the test, e.g. $p = 0.05$, must also be stated.

Example: A biscuit manufacturer received customer complaints reporting an 'off' flavour in a particular batch and wanted to determine whether there was a difference between that batch and standard production. They decided to run a duo-trio test with the objective to determine if a significant difference existed between the complaint batch and a standard batch of biscuits that had been manufactured at a similar time. As there was a limited amount of the complaint batch available, the test was conducted using the standard batch as a constant reference. The significance level chosen for the test was 5%. As recommended in the ISO standard, a panel of 32 untrained assessors participated in the duo-trio test on the two batches of biscuit, of which 17 assessors correctly identified the noncomplaint sample as being most similar to the reference.

From the table in Appendix 4, for a panel of 32 assessors, the minimum number of correct responses required at 5% significance level ($p = 0.05$) is 22. The test result does not exceed this value.

From software packages, the probability of making a type I error with this result is $p = 0.43$. This is greater than $p = 0.05$ (the significance level of the test).

Conclusion: There is no significant difference between the two batches of biscuit ($p > 0.05$). There is no evidence from this test to suggest that the 'complaint' batch had developed an 'off' flavour. The cause of the complaint could not, therefore, be linked to manufacture. It may, however, be due to other factors, e.g. storage conditions or damage to packaging in the supply chain.

5.2.2.3 Difference from control test

Objective: To determine if a difference exists between one or more samples and a control sample, and to determine the size of the difference between the sample(s) and the control.

Procedure: Assessors are presented with the control sample and a blind coded test sample. They are asked to assess the two samples and determine if a difference exists between them. They are provided with a scale to record the magnitude of the difference. Appropriate palate cleansers should be used after each sample.

The difference from control test is particularly useful for assessing samples that are not homogeneous. It can be used as a two-sample test when the samples are fatiguing or have significant carry-over. It is most commonly used for quality control where the assessors are trained to understand the scale and the typical variation in production samples. Training is critical so that assessors understand the relative distance of all the points along the scale in relation to production differences. In quality control, the application of this scale is most useful when only a few sensory characteristics vary during production. If samples vary in several attributes, the use of more specific attribute scales, rather than one overall difference scale, is more appropriate. For data interpretation, cut-off or action points will be applied to the scale at specific degrees of difference, e.g. pass/fail, rework or reject. Assessors should not be aware of these cut-off values, although in practice this may be difficult if QA/QC panels are part of the production team.

Experimental design: The control sample is presented first. In QA/QC, however, the control (target/standard product) is often well known through familiarity and training so that a remembered (or mental) control is sufficient. One or more test samples can be presented simultaneously. The test samples should include one or more blind coded control samples. To avoid fatigue, large numbers of test samples should not be presented in one session but split over a reasonable number of sessions. The presentation order for samples should be balanced.

Typically 20–50 subjects are required to determine the degree of difference. When this method is integrated into a QA/QC procedure, the number of highly trained assessors may be as few as five.

Questionnaire: See Figure 5.3.

Difference from control test

Assessor: Date: Sample code:

You are provided with a control sample and a test sample labelled with a three-digit code. Remove the lid and assess the aroma of the samples. Determine if the test sample is different to the control and record the magnitude of that difference on the scale below (please tick).

No difference	
Very slight difference	
Slight/moderate difference	
Moderate difference	
Moderate/large difference	
Large difference	
Very large difference	

Comments:

Figure 5.3 Example of a questionnaire for the 'difference from control' test.

Note: For this scale, the response is converted to a number between 1 and 7. Alternative numeric scales are commonly used, e.g. a numeric category scale from 0 to 9 where 0 is no difference and 9 is very large difference.

Data analysis: The mean score for each test sample and the blind control sample(s) is calculated. Difference scores for the coded control samples represent the degree of heterogeneity in the samples and/or simply the effect of asking the 'difference' question, i.e. this represents the placebo effect and serves as an experimental control for noise.

The raw data are analysed using two-factor ANOVA (if the data are normally distributed) (see Appendix 5). If a significant difference exists between the samples, a Dunnett's multiple comparison test (MCT) is used to determine which samples are significantly different to the control. Dunnett's test is a specialised MCT used for comparisons against a control. Other MCTs, e.g. Fishers LSD, can be used to determine if significant differences exist between the test samples. The calculations can be carried out by hand; however, it is more common to use statistical software packages to complete the analysis (see Appendix 5).

It is not good practice for ANOVA to be used if the data are not normally distributed as may be the case with naive assessors. In this instance, the data can be converted to ranks and analysed using Friedman's ANOVA for ranked data and Fishers least significant difference (LSD) for ranks (see Section 5.2.3.3).

Conclusion: From a 'difference from control' test, the conclusion would be that a significant difference does OR does not exist between the test samples and the control. The significance level of the test, e.g. $p = 0.05$, must also be stated. It is also possible to comment on the magnitude of the difference between sample(s) and control.

Example: A personal care company manufacturing shampoos ran a QC programme. This programme included difference from control testing to ensure the fragrance of the shampoo remained consistent. The objective of the difference from control testing was to determine if a significant difference existed between the aroma of samples from fragranced batches of shampoo and the fragranced control at 5% level of significance. In this instance, a panel of 35 assessors participated in the test. Four production samples and two blind coded control samples were compared to the control. Samples were presented simultaneously over two sessions, with three samples presented per session. The difference was rated on a numeric scale 0 (no difference) to 9 (extremely different). The mean panel data are shown in the following table.

Sample	Mean
Test 1	0.2
Test 2	6.1
Test 3	1.9
Test 4	0.4
Control 1	0.3
Control 2	0.1

The data were analysed using ANOVA (two-factor without replication) with samples and assessors as factors.

Factor	DF	SS	MS	F	p-value
Total	245	1403.51			
Sample	6	991.86	165.31	11950.06	<0.0001
Assessor	34	0.469	0.14	0.997	0.480
Error	204	3.502	0.14		

The ANOVA table shows a significant sample effect. (There is no significant assessor effect.)

The Dunnett's MCT gave the following results (where MS_{Error} is Mean Square Error).

$$\text{Dunnett's MCT range} = D((\sqrt{2} \times MS_{Error})/n)$$

From tables (see O'Mahony 1986), $D = 2.53$ (two-tailed alternative hypothesis; $df_E = 204$; significance level = 5%; number of comparisons = 6, including controls).

$$\text{Dunnett's MCT range} = 2.53((\sqrt{2} \times 0.14)/35) = 0.23$$

Therefore, mean scores must differ by more than 0.23 when compared to the blind control sample before a significant difference can be concluded.

Conclusion: Tests 2 and 3 were significantly different to both control samples; test 1 was not significantly different to either control; test 4 was significantly different to control 2 but not control 1. These results provided evidence of batch to batch variation in aroma of the shampoo. The company had problems of consistency during production that would need to be investigated further.

5.2.2.4 Same–different test

Objective: To determine if a difference exists between two samples.

Procedure: Assessors are presented with a pair of samples and asked to determine if the samples are the 'same' or 'different'; they may also be asked to describe any differences. Samples are labelled with three-digit codes (blind coded) and should be assessed in the order provided (left to right). Appropriate palate cleansers should be used after each sample.

The same–different test is useful when triangle and duo-trio tests are not suitable, e.g. when samples are too complex, when there is too much carry-over to present samples multiple times or when personal care products are to be assessed in half-head or half-face trials.

The same–different test is subject to response bias due to variation in assessors' criteria for assigning a sample as 'same' or 'different' (O'Mahony 1992). In order to minimise this bias, a sureness rating can be added to the test. In this instance, assessors are asked to indicate how sure they are about their decision, using a simple category scale, e.g. very sure, sure, unsure and very unsure.

Experimental design: There are four possible sample presentations. They are

AA   BB
AB   BA

Assessors either receive one, two or all four pairs. If the samples are complex, or need comparing in half-head or half-face trials, each assessor will receive only one pair. In this instance, all four possible presentations are used an equal number of times. Alternatively, each assessor may receive one 'same' pair and one 'different' pair, or all four possible pairs.

The same–different test is often used with 30–50 assessors, although this number may be increased to as many as 200, particularly when only one pair is given.

Questionnaire: See Figure 5.4.

Data analysis: The total number of responses for 'same' and 'different' are tallied for each sample presentation. The chi-squared test (χ^2) is used to compare sample presentations that are the same (AA and BB) with those that are different (AB and BA).

When calculating by hand, the χ^2 statistic is compared to a statistical table (see Appendix 6) that shows the minimum value required before it can be concluded that a significant difference exists between the samples. The significance level (typically 5%) must also be specified. Alternatively,

78 Sensory evaluation

Same–different test

Assessor:　　　　　　　Date:　　　　　　Sample codes:

You are provided with two samples, each labelled with a three-digit code. Assess each sample in the order provided, from left to right, and determine whether the samples are the 'same' or 'different'. Record your result below.

Cleanse your palate with cracker and water after each sample. You are not permitted to retaste the samples. Comment on any differences you experienced.

Samples are same	
Sample are different	

Comments:

Figure 5.4 Example of a questionnaire for the same–different test.

software packages provide not only the χ^2 statistic and the critical minimum value that must be exceeded, but also the probability of making a type I error should it be concluded that a significant difference exists between the samples.

If a sureness rating was used, the results from this test can also be analysed using a more complex data analysis technique known as R index. The R index provides a measure of discrimination between the products based on the theoretical number of 'correct' results, had the samples been presented as paired comparisons. More information on R index can be found in Appendix 11.

Conclusion: From a same–different test, the conclusion is that a significant difference does OR does not exist between the two samples. In either case, the significance level of the test, e.g. $p = 0.05$, must also be stated. In addition, a comment may be made about the nature of any detectable difference.

Example: A cosmetics company wished to determine whether a new, cheaper processing method could be substituted for the current process. The action standard for changing to the new process was no significant sensory difference at the 5% level between the two processes.

The company decided to run a same–different test with the objective of determining if a significant difference existed between the two cream samples. The significance level of the test was chosen to be 5%. A panel of 120 assessors participated in the test on two samples of unfragranced and uncoloured face cream made from the same ingredients but from two different processing methods. Each assessor received one pair of samples (matched or unmatched) and was asked to assess their textural properties by using each cream on one half of his/her face and determining if the samples were the 'same' or 'different'. The presentation order was balanced for sample pair and order within each pair. The results are summarised as follows.

Subjects responded	Subjects received		Total
	Matched pairs (AA or BB)	Unmatched pairs (AB or BA)	
Same	37	18	55
Different	23	42	65
Total	60	60	120

$$\chi^2 = \Sigma((\text{Observed} - \text{Expected})^2/\text{Expected})$$

Observed values are the assessor responses and expected values are calculated for each option of 'subjects responded/subjects received', i.e. same/matched, same/unmatched, different/matched and different/unmatched.

Expected values (E) were calculated as follows.

same/matched $E = 55 \times 60/120 = 27.5$

same/unmatched $E = 55 \times 60/120 = 27.5$

different/matched $E = 65 \times 60/120 = 32.5$

different/unmatched $E = 65 \times 60/120 = 32.5$

$$\chi^2 = \left(\frac{(37-27.5)^2}{27.5}\right) + \left(\frac{(18-27.5)^2}{27.5}\right) + \left(\frac{(23-32.5)^2}{32.5}\right) + \left(\frac{(42-32.5)^2}{32.5}\right)$$
$$= 12.1$$

From the table in Appendix 6, the critical value for the chi-squared test (χ^2) is 3.84 ($n - 1$ degrees of freedom, $\alpha = 0.05$). The calculated χ^2 statistic (12.1) exceeds this value, indicating that a significant difference exists between the two samples.

Conclusion: There is a significant difference between the two samples of face cream when applied to the face ($p < 0.05$). Assessors' comments suggested that sample A was thicker and tackier when applied to the skin. The action standard was not met and the new process was not adopted.

5.2.2.5 'A' 'not A' test

Objective: To determine if a difference exists between two samples.

Procedure: Initially, assessors are presented with two samples, 'A' and 'not A', and asked to familiarise themselves with their characteristics. The samples must be labelled appropriately, e.g. control/not control, target/not target and standard/not standard. Alternatively, assessors may be presented with a range of samples that represent the typical variation in 'A' and 'not A'. Assessors are usually given as much time as necessary to familiarise themselves with the samples. These are then removed and the assessors are presented with a series of individual samples, labelled with random three-digit codes, and asked to determine if they are the same as 'A' or 'not A'. Appropriate palate cleansers should be used after each sample.

Similar to the same–different test, 'A' 'not A' is used when the triangle and duo-trio tests are not suitable, e.g. when samples are too complex, when there is too much carry-over to present samples multiple times or when personal care products are to be assessed in half-head or half-face trials. 'A' 'not A' is used in preference to the same–different test when one of the samples has specific meaning or is well known to the panel, e.g. reference or control.

The 'A' 'not A' test is subject to response bias due to variation in assessors' criteria for assigning a sample as 'A' or 'not A' (O'Mahony 1992). In order to minimise this bias, a sureness rating can be added to the test. In this instance, assessors are asked to indicate how sure they are about their decision, using a simple category scale, e.g. very sure, sure, unsure and very unsure.

Experimental design: Usually, 10–50 assessors are trained to identify the 'A' and 'not A' samples. During the test, assessors receive either:
- one sample (either 'A' or 'not A');
- two samples ('A' and 'not A');
- several samples (up to 20 samples, equal numbers of 'A' and 'not A').

The number depends on the amount of carry-over and fatigue associated with the samples. When several samples are presented, the presentation order should be at least randomised and, if possible, balanced, and the results recorded on separate questionnaires to avoid assessors looking for patterns in the data.

The most common design involves only one 'A' sample and one 'not A' sample; however, it is possible to modify this test to include 2–3 different 'not A' samples, all of which must be presented in the initial familiarisation. When 'A' 'not A' tests are used in a QC programme, 'not A' samples may be unknown and unavailable for familiarisation.

Questionnaire: See Figure 5.5.

Data analysis: The total number of responses for 'A' and 'not A' are tallied for each sample presentation. The chi-squared test (χ^2) is used to compare the different sample presentations and their responses.

When calculating by hand, the χ^2 statistic is compared to a statistical table (see Appendix 6) that shows the minimum value required before it can be concluded that a significant difference exists between the samples. The significance level (typically 5%) must also be specified. Alternatively, software packages provide not only the χ^2 statistic and the critical minimum value that must be exceeded, but also the probability of making a type I error (p-value) should it be concluded that a significant difference exists between the samples. This analysis is not wholly appropriate for the design involving multiple sample presentations to each assessor; however, it is commonly used and the p-value is considered to be a good approximation.

'A' 'not A' test

Assessor: Date:

You are provided with two samples, each labelled with a three-digit code. The samples are either 'A' or 'not A' as experienced in the preliminary session. Assess each sample in the order provided, from left to right, and determine its identity. Record your result below.

Cleanse your palate with cracker and water after each sample. You are not permitted to retaste the samples.

Sample	A	Not A
219		
470		

Comments:

Figure 5.5 Example of a questionnaire for the 'A' 'not A' test.

82 Sensory evaluation

If a sureness rating is used, the results from this test can also be analysed using the R index procedure. The R index provides a measure of discrimination between the products based on the theoretical number of 'correct' results should the samples be presented as paired comparisons. For example, if 10× 'A' samples and 10× 'not A' samples were presented to an assessor, this would be equivalent to 100 paired comparisons (each 'A' compared to each 'not A'). For more information on the use of R index and its advantages, see Appendix 11.

Conclusion: From an 'A' 'not A' test, the conclusion is that a significant difference does OR does not exist between the two samples. In either case, the significance level of the test, e.g. $p = 0.05$, must also be stated.

Example: A food manufacturer wanted to change supplier of the milk that is used as an ingredient in one of its products. The action standard for making the change was no significant difference at the 5% level between milk from the old and potential suppliers. The 'A' 'not A' test was selected with the objective of determining if a significant difference existed between milk samples from the two suppliers. A panel of 50 assessors participated in the 'A' 'not A' test on two samples of milk, one from each supplier. Each assessor was familiarised with the sensory characteristics of the target sample ('A') from the old supplier and the nontarget sample ('not A') from the new supplier, and then received one test sample and asked to identify it as the 'target' or 'not the target'. The significance level of the test was 5%. The results are summarised as follows.

		Subjects received		Total
		A	Not A	
Subjects	A	34	20	54
responded	Not A	16	30	46
Total		50	50	100

$$\chi^2 = \Sigma((\text{Observed} - \text{Expected})^2/\text{Expected})$$

Observed and expected values were determined for each responded/received option (A/A; A/not A; not A/A and not A/not A). Expected values were calculated as follows.

A/A $\qquad E = 54 \times 50/100 = 27.0$

A/not A $\qquad E = 54 \times 50/100 = 27.0$

Not A/A $E = 46 \times 50/100 = 23.0$

Not A/Not A $E = 46 \times 50/100 = 23.0$

$$\chi^2 = \left(\frac{(34-27.0)^2}{27.0}\right) + \left(\frac{(20-27.0)^2}{27.0}\right) + \left(\frac{(16-23.0)^2}{23.0}\right) + \left(\frac{(30-23.0)^2}{23.0}\right)$$
$$= 7.9$$

From the table in Appendix 6, the critical value for the chi-squared test (χ^2) is 3.84 ($n - 1$ degrees of freedom, $\alpha = 0.05$). The calculated χ^2 statistic (7.9) exceeds this value indicating a significant difference between the two samples.

Conclusion: There is a significant difference between the two samples of milk ($p < 0.05$). The action standard was not met and the change was not made at this stage. A suitable next step, however, may be to determine if there is any significant difference between the final products made with the two milk samples and/or which sample is preferred.

5.2.3 Attribute-specific tests

For attribute-specific tests, assessors are directed to focus on one specified attribute or quality.

5.2.3.1 Paired comparison (2-AFC)

Objective: To determine if a difference exists between two samples with regard to a specified attribute, e.g. sweetness, hardness and intensity of fragrance.

Procedure: Assessors are presented with two blind coded samples. They are asked to assess the samples and determine which of the two has the greatest intensity of a specified attribute. Assessors may be pretrained on the attribute, depending on the test objectives. Appropriate palate cleansers should be used after each sample. Ideally, the samples should vary only in intensity of the attribute in question, although practically this is very hard to achieve. If there are too many differences between samples, an overall discrimination test should be used, e.g. triangle test.

The paired comparison test is rapid and easy to use. It can also be used for assessing preference between two samples, in which case it is referred to as a paired preference test and the question asked is which sample is preferred (see Section 5.4.5.1).

84 Sensory evaluation

Experimental design: Samples are presented in pairs. There are two possible orders of presentation which should be used an equal number of times. They are

AB
BA

A minimum of 30 assessors should be used, although some texts vary in terms of their recommendation.

Questionnaire: See Figure 5.6.

Data analysis: Determine the total number of times each sample is selected. There are two ways of analysing the data.

When calculating by hand, the larger number of responses for one sample is compared to statistical tables (see Appendix 7). The table states the *minimum* number of responses required before a significant difference can be concluded from the test. The significance level of the test must also be specified (typically 5%).

Alternatively, software packages calculate the probability of making a type I error should it be concluded that a significant difference exists between the samples.

Conclusion: From a paired comparison test, the conclusion is either that one sample is significantly more intense than the other with regard to the specified attribute, or that there is no significant difference between them with regard to the specified attribute. The significance level of the test, e.g. $p = 0.05$, must also be stated.

Paired comparison test

Assessor: Date:

You are provided with two samples of toilet paper. Please assess each sample for softness and determine which is the softest. To assess softness rub the sample between your thumb and index finger. Record your result below.

Sample	Softest (please tick)
297	
831	

Comments:

Figure 5.6 Example of a questionnaire for the paired comparison test.

Sensory test methods 85

> **Example:** A paper products company wanted to compare the softness of their toilet paper with that of their biggest competitor. A panel of 60 assessors was trained to assess and identify softness in a consistent manner by rubbing between forefinger and thumb. The panel participated in a paired comparison test to determine if there was a significant difference in softness between the two samples of toilet paper. The results showed that 47 out of the 60 assessors selected sample A, the company product, as the softest.
>
> From the table in Appendix 7, for a panel of 60, the minimum number of identical responses required to determine that a difference exists at 5% significance level ($p = 0.05$) is 39.
>
> From software packages, the probability of making a type I error with this result is $p = 0.0009$. This is less than $p = 0.05$ (the significance level of the test); it is also less than a 0.01% level of significance.
>
> *Conclusion:* There is a significant difference in softness between the two samples of toilet tissue – sample A, the company product, is significantly softer than sample B, the competitor product ($p < 0.05$). On the basis of these results, the company subsequently carried out a further test to substantiate an advertising claim that their toilet tissue was softer than other leading brands.

5.2.3.2 3-Alternative forced choice

Objective: To determine if a difference exists between two samples with regard to a specified attribute, e.g. sweetness, hardness and intensity of fragrance.

Procedure: Assessors are presented with three blind coded samples. Two samples are the same and one is different, although the assessor is not made aware of this fact. They are asked to assess the samples in the order provided and determine which sample has the highest intensity of a specified 'attribute'. Assessors may be pretrained on the attribute, depending on the test objectives. Appropriate palate cleansers should be used after each sample.

As with the 2-alternative forced choice (2-AFC) test, samples should vary only in intensity of the attribute in question, although practically this is very hard to achieve. If there are too many differences between samples, overall discrimination tests should be used, e.g. triangle test.

This method is commonly used to determine threshold values, i.e. the lowest concentration of a compound that can be detected, whereby the 'same' samples are the diluent or carrier (water, air) and the 'different' sample contains the stimulus in the diluent or carrier (see ISO 13301:2002).

86 Sensory evaluation

Experimental design: There are only three possible orders of sample presentation. They are

AAB
ABA
BAA

It is good practice to use each possible presentation order an equal number of times with a minimum of 24 assessors, although the absolute number chosen depends on the overall aim and the significance level selected. Typically, the sample assumed to be the most intense is presented as the 'odd' sample; however, when the most intense sample cannot be predicted, the test may need to be performed twice with each sample being presented as the 'odd' sample.

Questionnaire: See Figure 5.7.

Data analysis: Determine the total number of times the 'odd' sample is selected. There are two ways of analysing the data.

When calculating by hand, the number of 'correct' responses is compared to statistical tables (see Appendix 3). The table states the *mini-*

Sample test: lavender aroma

Assessor: Date:

You are provided with three shower gel samples, each labelled with a three-digit code. Assess each sample in the order provided, from left to right, and determine which sample has the most intense lavender aroma. Record your result below.

Do not sniff the samples too vigorously and leave 10 seconds between samples to give your nose a chance to recover. You may resniff the samples.

DO NOT CONSUME THE SAMPLES

Sample	Most intense aroma (please tick)
219	
470	
593	

Comments:

Figure 5.7 Example of a questionnaire for the 3-AFC test. (*Note*: The ballot does not use the name of the test to avoid providing too much information and biasing the assessors.)

mum number of 'correct' responses before a significant difference can be concluded from the test. The significance level of the test must also be specified (typically 5%).

Alternatively, software packages calculate the probability of making a type I error should it be concluded that a significant difference exists between the samples.

Conclusion: From a 3-alternative forced choice (3-AFC) test, the conclusion is either that one sample was significantly more intense than the other with regard to the specified attribute or that there was no significant difference between them with regard to the specified attribute. The significance level of the test, e.g. $p = 0.05$, must also be stated.

> **Example:** A company manufacturing personal care products had improved the lavender fragrance in their shower gel and wanted to determine whether the same concentration of the new fragrance gave a similar perceived intensity of lavender aroma when compared to the old fragrance. They chose to run a 3-AFC test with the objective to determine if there was a significant difference in the intensity of lavender aroma between two samples of shower gel (samples A and B). A panel of 30 assessors participated in the 3-AFC test. The direction of the response could not be predicted, so the test was performed twice, once with sample A as the 'odd' sample and once with sample B as the 'odd' sample. The results showed that in the first test, 9 out of the 30 assessors selected sample A as the most intense and in the second test, 11 out of 30 assessors selected sample B as the most intense.
>
> From the table in Appendix 3, for a panel of 30, the minimum number of identical responses required at 5% significance level ($p = 0.05$) is 15. This is greater than the result for either of the 3-AFC tests.
>
> From software packages, the probability of making a type I error with this result is $p = 0.71$ for 'A' as the odd sample, and $p = 0.42$ for 'B' as the odd sample. In both instances, this is greater than $p = 0.05$ (the significance level of the test).
>
> *Conclusion:* There is no significant difference in intensity of lavender aroma between the two samples of shower gel ($p > 0.05$). The company used the same concentration of the new fragrance as they had for the old fragrance.

5.2.3.3 Ranking test

Objective: To determine if a difference exists between three or more samples with regard to a specified attribute, e.g. sweetness, hardness and intensity of fragrance.

88 Sensory evaluation

Procedure: Assessors are presented with several blind coded samples. They are asked to assess the samples in the order provided and place them in order of intensity for a specified attribute. Assessors may be pretrained on the attribute depending on the test objectives. Typically, assessors are forced to make a choice for each ranking position; however, it is possible to allow ties between samples. Appropriate palate cleansers should be used after each sample. Ranking is particularly useful for sorting samples prior to additional analysis and also for descriptive panel training. Rank data can be subject to R index analysis (see Section 5.2.2.5 and Appendix 11).

Experimental design: The number of samples assessed in a ranking test depends on how fatiguing their assessment will be to the assessors. For example, it is possible to use as many as 8–10 samples for simple products such as mineral water, or when assessing attributes that do not require consumption. More typically, 5–6 samples are used when the intensity of taste/flavour attributes is being ranked.

The order of sample presentation should be balanced across the panel such that each sample is assessed in every possible position an equal number of times (see Section 3.7.2)

Questionnaire: See Figure 5.8.

Ranking test

Assessor: Date:

You are provided with five samples of peanut butter, each labelled with a three-digit code. Please assess the samples in the order provided and place the samples in increasing order of saltiness. Record your result below.

Please make sure that you cleanse your palate after samples.

Salt intensity	Sample number
1st = least salty	
2nd	
3rd	
4th	
5th = Most salty	

Comments:

Figure 5.8 Example of a questionnaire for the ranking test.

Sensory test methods 89

Data analysis: The data are summarised in a table showing rank order for each assessor. When ties in rank order are permitted the data must be modified before analysis. The available rank orders are summed and divided by the number of samples tied for that position. For example, in a four product ranking test, if samples are ranked as first, second and tied for third (most intense), the rank order of the two tied samples is $(3 + 4)/2 = 3.5$; rank orders would be entered as 1, 2, 3.5 and 3.5. The rank orders are summed to produce rank sums for each product (see following example). The Friedman statistic (T) is then calculated. Note that the calculation differs for tests allowing tied ranks.

When calculating by hand, the T statistic is compared to a statistical table (see Appendix 8) that shows the minimum value required before it can be concluded that a significant difference exists between two or more of the samples. The significance level (typically 5%) must also be specified. Alternatively, software packages provide not only the T statistic and the critical minimum value that must be exceeded, but also the probability of making a type I error should it be concluded that a significant difference exists between the samples.

If the Friedman analysis shows that a significant difference exists between two or more samples, the identity of the different samples is determined using Fishers least significant difference multiple comparison test for ranks (LSRD) used at the same significance level (5%). The LSRD formula calculates a value that is compared to the difference between rank sums. If the difference exceeds the LSRD value, the samples are said to be significantly different.

Conclusion: In a ranking test, the conclusion is either that no significant difference exists between the samples or that a significant difference exists between specified samples; these are usually listed. The ranked attribute and the significance level of the test, e.g. $p = 0.05$, must also be stated.

Example: A soft drinks manufacturer decided to carry out a market assessment of fizzyness in carbonated lemonade. They decided to carry out a ranking test to determine if there were significant differences in fizzyness between the four leading brands. A panel of 15 assessors participated in a ranking test of 4 samples of lemonade (E-H). Table 5.1 summarises the results.
  The T statistic is calculated as follows.

$$T = (12\Sigma R^2/bt(t + 1)) - (3b(t + 1))$$

where t is the number of samples, b the number of assessors and R the rank sum.

Sensory evaluation

Table 5.1 Rank order of fizzyness for 15 assessors ranking four products, and resulting overall rank sums

Assessor	E	F	G	H
1	1	3	2	4
2	1	2	3	4
3	1	2	4	3
4	2	1	3	4
5	1	3	2	4
6	3	1	2	4
7	1	3	2	4
8	1	3	2	4
9	3	2	1	4
10	1	3	4	2
11	1	2	3	4
12	1	2	3	4
13	1	2	4	3
14	3	1	2	4
15	1	3	2	4
Rank sum	**22**	**33**	**39**	**56**

$$T = (74760/300) - 225 = 24.2$$

From the table in Appendix 8, the critical value for the Friedman test is 7.81 ($n - 1$ degrees of freedom, $\alpha = 0.05$). The calculated T statistic (24.2) exceeds this value and, therefore, Fishers LSRD must be used to determine which samples are significantly different.

Fishers LSRD ($\alpha = 0.05$) is calculated as follows.

$$\text{LSRD} = t_{\alpha/2\infty}\sqrt{(bt(t + 1)/6)} = 1.96\sqrt{50} = 13.9$$

Where, $t_{\alpha/2}$ is taken tables for Student's t distribution (see O Mahoney 1986)

Samples whose rank sums differ by more than 13.9 are deemed to be significantly different. The results are summarised in Table 5.2.

Table 5.2 Summary of ranking results for four lemonade products

Sample	Rank sum	Significance[a]
H	56	A
G	39	B
F	33	BC
E	22	C

[a]Samples sharing the same letter are not significantly different ($p < 0.05$).

> From software packages, analysis of the same data gives the probability of making a type I error as $p \leq 0.0001$. This is less than $p = 0.05$ (the significance level of the test); it is also less than a 0.01% level of significance.
> *Conclusion*: There is a significant difference in fizzyness between the four samples ($p < 0.05$). Sample H is significantly more fizzy than all other samples; samples G and F are not significantly different and neither are samples F and E. Sample G is significantly more fizzy than sample E. The company was able to conclude that there were significant differences in fizzyness between the four leading brands of carbonated lemonade. On the basis of the test, they decided to carry out a follow-up study to determine the level of fizzyness that is preferred by consumers.

5.2.4 Similarity

Some of the methodologies described earlier can be used to determine the degree of similarity between products. In fact, many of the objectives for running a discrimination test are truly about similarity and not difference, e.g. the need to change ingredients or any aspect of processing without changing the sensory characteristics of the product. It is wrong to assume that no significant difference also means that the products are similar.

For discrimination, it is important to reduce the risk of saying that samples are different when in fact they are not. This is an example of a type I error and it is minimised by reducing the significance level or α risk of the test, typically to a value of 5%, although in some circumstances, this can be higher or lower. For similarity, it is important to reduce the risk of saying that samples are not different when, in fact, they are. This is an example of a type II error and it is minimised by increasing the statistical power of the test and, therefore, reducing the β risk. In practical terms, this is achieved by increasing the number of participants, setting a reasonable level for people who can truly discriminate between the samples (Pd) and allowing the α risk to become much larger. These concepts are described in more detail in the following sections.

5.2.4.1 The power of the test
Power is related to the risk of making a type II error (β); power $= 1 - \beta$.

If β (the risk of saying the samples are not significantly different when they are) is decreased, the power of the test increases. This increase in power means that differences are more likely to be found if they exist. When the objective of a discrimination test is to determine the degree of similarity, then increasing power is very important so that any differences that do exist will not be missed. Ironically, the power of the test is often ignored when the objective for the test is to determine differences between samples; in this case type I error (α risk) is minimised so that differences are not reported in error. Increasing the number of participants increases the power of the test.

5.2.4.2 Proportion of true discriminators

In a discrimination test, it is likely that some assessors will be able to truly tell a difference between the samples. This group of individuals are 'the proportion of true discriminators (Pd)'. The number of 'correct' responses in a test, therefore, includes a proportion of people who could truly tell the difference (Pd) and the remainder who correctly guessed the answer. When testing for similarity, the value of Pd has a direct effect on the number of assessors you need to use at a specified level of β. Whilst it may seem sensible to reduce the Pd to <5%, in practical terms, this will require the use of several hundred assessors to keep the β-risk below 10% (power = 90%). It is common practice to consider three levels for Pd: low (<25%), medium (25–35%) and high (>35%).

5.2.4.3 Selecting the correct number of assessors

The requirement to minimise β and increase power, increases the number of assessors that will be needed. As mentioned earlier, this can be reduced to a practical level by allowing a higher level of true discriminators (Pd). However, this may not be commercially advantageous and the project may require a much smaller level of Pd, e.g. forced ingredient or process changes in leading brands. Another way of minimising the number of assessors is to allow the risk of a type I error (α) to become much larger (>20%). This becomes possible as it would not normally be expected to find significant differences between samples that are being assessed for similarity. Of course, there may be occasions when α, β and Pd need to be minimised, in which case the large sample numbers may make the test impossible.

To determine how many assessors should be used in a similarity test, tables and software packages list numbers of assessors for set values of α, β and Pd (see ISO standards for individual test methods).

5.2.4.4 Using the triangle test to determine similarity

Triangle and duo-trio tests are the most commonly used tests to determine similarity. In practical terms, the procedure is conducted as described in Sections 5.2.2.1 and 5.2.2.2, respectively. The only difference is in the number of assessors used to conduct the test and the method of interpreting the data. The following details how to use a triangle test to determine similarity.

Objective: To determine the degree of similarity between two samples.

Procedure: Assessors are presented with three samples and told that two samples are the same and one is different. They are asked to assess the samples in the order provided and determine which sample is 'the odd one out'. They may also be asked to describe the difference. Appropriate palate cleansers should be used after each sample. Samples are labelled with three-digit codes (blind coded).

Note: This is the same procedure as that used to assess product differences.

Experimental design: There are six possible orders of sample presentation. They are

AAB BBA
ABA BAB
BAA ABB

In some instances, only one half of the design is used, for example if the quantity of one of the samples is limited, or if one of the samples is the standard/reference and, therefore, is presented as the duplicate sample. It is good practice to use each possible presentation order an equal number of times with a minimum of 60 assessors; the maximum number will vary depending on the levels of α, β and Pd set for the test.

Questionnaire: See Figure 5.9.

Data analysis: The total number of responses correctly identifying the 'odd' sample is counted. There are two ways of analysing the data.

If analysing the data by hand, compare the number of correct responses to statistical tables (see Appendix 3). Determine the *minimum* number of correct identifications required before a significant difference can be concluded at the significance level (α) set for this test. In this instance, the objective is to determine the degree of similarity so a significant difference is not expected, even at α levels up to 20%. If the total number of 'correct' responses does not exceed the minimum, a statement can be made about similarity that must include reference to the β risk and the proportion of true discriminators (Pd) set for the test.

Alternatively, software packages calculate the probability of making a type I error (α risk), and a type II error (β risk) at low, medium and high levels of Pd. In some cases, they provide values for Pd at low (1%), medium (5%) and high (10%) levels of β. From the output, it is possible to make a statement about similarity that fits the objectives of the test

94 Sensory evaluation

Triangle test

Assessor: Date:

You are provided with three samples, each labelled with a three-digit code. Two samples are the same and one is different. Assess each sample in the order provided, from left to right, and select the 'odd' sample. Record your result below.

Cleanse your palate with cracker and water after each sample. You are not permitted to retaste the samples. Please say why the odd sample is different. Comment on any differences you experienced.

Sample	Different sample (please tick)
219	
470	
593	

Comments:

Figure 5.9 Example of a questionnaire for the triangle test.

and the levels of α, β and Pd considered appropriate. The large amount of information provided in software output allows the user to be more focused on the similarity statement.

Conclusion: When used for similarity, the conclusion is that no more than $x\%$ of a population (Pd) can truly tell the difference between the samples ($\beta = y\%$). The level of β can also be stated as a level of confidence in the result, for example if the level of β was set at 1%, the conclusion may state that with a 99% level of confidence, no more than $x\%$ of the population can truly tell a difference between the samples.

Example: A raw material is in short supply and an alternative has been sourced. It is imperative that the consumer cannot tell a difference between the original product and that made with the new ingredient. Tables stated that at $\beta = 1\%$, $\alpha = 20\%$ and Pd = 20%, 140 assessors must be used for the test. Forty-five assessors correctly identified the odd sample.

From the table in Appendix 3, for a panel of 140, the minimum number of correct responses required at 20% significance level ($p = 0.20$) is 53.

The test result does *not* exceed this value. A statement can be made about the similarity of the samples.

From software packages, the probability of making a type I error with this result is $p = 0.65$. The output relating to similarity is as follows.

Pd (%)	β risk	Power
12	0.01	0.99
15	<0.01	>0.99
20	<0.01	>0.99
27.5	<0.01	>0.99
35	<0.001	>0.999

Conclusion: When analysing the data by hand, there is 99% confidence that no more than 20% of the population can truly tell a difference between the products. Note that the statement can only refer to the parameters that were predefined at the start of the test (Pd, α and β).

When using the software output, there is 99% confidence that no more than 12% of the population can truly tell a difference between the products, or there is >99% confidence that no more than 20% of the population can truly tell a difference between the products. Note that software provides a great deal of information relating to the actual results collected and a more focused statement can be made.

The company concludes that the products are sufficiently similar to allow the new raw material to be used; they are confident (99%) that only a small proportion of the population (12%) can tell a difference between the products.

5.2.5 Dos and don'ts

- ✔ Select the correct method for the objective (difference or similarity; overall or attribute specific).
- ✔ Select the correct method for the sample type and quantity.
- ✔ Give clear written and verbal instructions to the panel.
- ✔ Provide equivalent portion sizes for each sample.
- ✘ Do not combine a discrimination test with a question regarding preference – it will bias both results.
- ✘ Do not replicate discrimination tests to get more responses and analyse the data as described earlier.

5.3 Descriptive analysis tests

5.3.1 Introduction

Descriptive analysis characterises the sensory properties of a product. Sensory qualities, their intensity and occurrence over time can be measured using this technique. A precise sensory description of a product can be generated and sensory differences between products can be described and quantified. Quantitative descriptive data can be linked to consumer data to understand sensory drivers of product liking, and linked to formula and instrumental measures to understand the chemical and physical components of a product that influence sensory characteristics.

5.3.1.1 Determining objectives and future needs

When setting up descriptive analysis, the following need to be considered in order to determine the method, type of assessors and training required:
- *Application*: How will the results be used? Applications include product development, product optimisation, market assessment, competitive assessment, QA/QC such as shelf life testing, etc., and these will determine the methodology to be used.
- *Product range*: What products will be assessed? These may be products of the same type with small variations in formulation, products from the same category and/or products from a diverse range of categories. This will determine the range of sensory space that needs to be covered and the sensitivity of the methodology, including number of replications and type of scale to be used. In general, products profiled in a single study should be of a similar type or generic group, e.g. different brands or development samples of the same flavour yoghurt. The number of products to be profiled may also impact the methodology chosen.
- *Statistical analysis*: How will the results be analysed to meet objectives? Different descriptive analysis methodologies lend themselves to different types of statistical analysis, which will impact the method selected.
- *Duration*: How long will descriptive analysis be required? This may be for one study, for many studies over the course of several years or for continuous tracking such as in QA/QC. This will determine the type of assessors and the training required.

5.3.1.2 Role of the panel leader

The role of the panel leader will range from passive facilitator (e.g. Quantitative Descriptive Analysis® (QDA®)) to directive leader (e.g. Spectrum™ method) depending on the descriptive methodology used. It requires skills and training to carry out the role effectively. The

characteristics and techniques most commonly employed by a panel leader include the following:
- Nonjudgemental approach
- Sensitive and assertive, yet diplomatic approach
- Active listener with the ability to probe for information
- Ability to handle diverse opinions and personalities
- Ability to motivate panel
- Recognises and guards against moderator bias
- Not opinionated, doesn't proffer information (except for Spectrum™ method)

5.3.2 Key steps of descriptive analysis

This section outlines the key generic steps in carrying out descriptive analysis: selection and general training of assessors, training of assessors for the study (generating attributes and assessment protocol, intensity calibration, performance check), evaluating samples, data analysis and reporting. These steps are common across most descriptive methods. There are, however, key differences and these are highlighted in Section 5.3.4. Descriptive analysis may also be used to produce a qualitative sensory profile by omitting the rating stage.

5.3.2.1 Selection and training of assessors

Descriptive analysis requires a small number of highly trained assessors. It is typically carried out with 6–18 assessors, who have been preselected to have good sensory abilities and received general training (see Sections 4.5.3 and 4.5.4). They are then trained (or calibrated) as described in the following sections. The degree of training, and the amount of experience gained in using the technique, will influence the variability of the data and, hence, the size of difference that can be detected by the panel.

5.3.2.2 Generating attributes and references

Attribute generation

In the first step of descriptive analysis, assessors are exposed to all samples, or at least a subset of samples that represent the extremes and illustrate all attributes. Assessors generate terms to describe the qualities of sensations present, or select attributes from a predefined list. The generated list is then refined so that it includes only objective, unique, unambiguous, independent sensory terms. Terms that are hedonic, e.g. 'nice', 'great', and consumer attributes, e.g. 'fresh', 'natural', should not be included. It may be desirable to include integrated attributes, such as 'total flavour intensity', but in general, combination attributes, such as 'creamy', should be broken down

98 Sensory evaluation

into their elemental parts, e.g. 'creamy texture' may be a combination of smooth, thick and oily. Including an 'other' option on the ballot allows sensory qualities not captured during this phase to be rated, and helps prevent dumping of uncaptured sensations into inappropriate attributes. The meaning of 'other' can be probed in subsequent discussions.

Agreement on attributes

In general, assessors agree the perceptual meaning of the attributes and produce a sensory lexicon of clearly defined terms. This includes an attribute name, written definition, method of assessment (e.g. biting with incisor, stroking with fingers) and physical reference(s) (food, nonfood, chemical compound, etc.) that illustrates the sensory experience of the attribute. It may happen that different terms are identified with the same sensory meaning (duplicate terms), in which case agreement should be reached on which term is to be used and the others eliminated.

5.3.2.3 Determining assessment protocol

During or after the generation and agreement of attributes, the product assessment protocol must be determined. This includes the way in which the product needs to be assessed in order to study each attribute, the point during the product assessment when each attribute will be assessed, the order in which attributes are assessed and methods to reset the senses back to a neutral state between samples.

The product assessment may be phased, e.g. for foods – smell, first bite, chew, swallow, aftertaste; for personal wash products – from the bottle/bar, on first lather in water, during use on skin, immediately after drying skin, several hours after use. Attribute order should be logical, e.g. aftertaste is assessed at the end.

The assessment protocol should control for bias (see Section 2.2). In some instances, the test objectives may require some adjustment to the protocol. For example, in order to independently assess flavour, colour differences may need to be masked.

Some products may require a long assessment, e.g. chewing gum may require a 30 minute assessment, and skin cream and deodorant may require assessment over several days, and some attributes may linger for a long time, e.g. chilli burn. It is important to allow a suitable time period over which to make assessments. It may be necessary to use time intensity (TI) methods, which are covered later in this section.

Some products may require a carrier, such as margarines and spreads. The carrier should be bland and of consistent quality, e.g. bread, crackers (see Section 4.4.2.3).

Sensory test methods 99

The assessment protocol should also include methods to overcome sensory fatigue and adaptation between samples, e.g. leaving a time gap between samples, and use of palate cleansers such as water, crackers, plain yoghurt, apple, cucumber and melon (See Section 4.4.2.5).

5.3.2.4 Rating intensity
Once the attributes and assessment protocol have been agreed, the product can be evaluated by rating the intensity of each attribute on a scale. There are several steps to this process depending on the method and type of scale used. For examples of different types of scale, see Appendix 9.

Scale design
Absolute and relative scales
For absolute scales, the range of intensities represented by the scale is equivalent in strength. This can be true across different attributes within a modality and for attribute scales used across different studies. These scales are used when it is important to make direct comparisons across attributes and studies.

In contrast, for relative scales, the range of the scales for different attributes and/or across different studies has different meanings in terms of strength. This approach is used to maximise sensitivity. For example, the sensory space covered by the scale is set to match the sensory space to be covered by the study. It is still possible to make comparisons across studies using common samples, common references and appropriate statistical techniques.

Intensity range covered by the scale
It is important to select an appropriate intensity range that will cover the perceived intensity range of the products to be assessed, whilst allowing for appropriate sensitivity of discrimination. Some examples are as follows:
- *Universal scale*: Covers the full range of sensations that may occur across all product classes and is useful when the panel is likely to assess a broad variety of products. It is said to be absolute. It is also said to be less sensitive to small differences.
- *Category-specific scale (product-specific scale)*: Covers the intensity range of a category/product class and is useful when the panel is going to work within one product category. It may be set up as an absolute or relative scale. It is said to be more sensitive than the universal scale.
- *Study-specific scale*: Covers the intensity range for an individual study. It may be set up as an absolute or relative scale. The relative version

of this scale is said to be the most sensitive methodology and is that typically described in text books as the classical descriptive analysis methodology.

Bipolar scale
The scale runs from one quality to another, rather than none to strong. It is advisable to avoid these scales as they give less information, but it may not always be possible. If bipolar scales are necessary, it is important to ensure that opposites are used, e.g. soft to hard is appropriate, but sour to sweet is not.

Scale labelling
The scale is labelled with numbers or words, e.g. 'weak', 'medium', 'strong'. Terms that imply hedonics, e.g. 'much too strong', should be avoided. Caution must be exercised as individual assessors may differ in their interpretation of labels unless they have been specifically trained to agree on their perceptual meaning, e.g. by using intensity references.

Training the panel to rate intensity
The depth of training will vary depending on the type of methodology and scale used. In unmodified QDA®, for example, the training is short and the aim is to ensure each assessor is consistent in their own scoring. In Spectrum™, training is intensive and can take up to 6 months as all assessors need to score on all scales in the same way. Intensity training can include the following:
- Basic training on how to use the scale type.
- Training to promote use of the ends of the scale.
- Calibration across assessors.
- Training to improve consistency and reproducibility so that assessors are consistent within themselves and/or with the rest of the panel, and are repeatable. Feedback should be given on performance and retraining given as necessary.

In some instances, intensity references (anchors) may be used to illustrate intensity points on the scale, such as, end-of-scale anchors to illustrate low and high ends of the scale and/or one or more mid-range intensity references. Examples include the following:
- One sample (standard) used to illustrate intensity across all attributes.
- One attribute, e.g. sweetness, saltiness, and so on, used to illustrate different points along the intensity scale via solutions at a range of different concentrations. This would be an absolute-type scale.

- Different intensity references for each attribute and/or different points on attribute scales.

A range of practical exercises can be used to train the panel on intensity. Examples include the following:

- Practise scaling by rating amount of shading on shapes (see Meilgaard *et al.* 2007).
- Early group discussions to reach consensus on intensity.
- Ranking or rank/rating of samples with a range of intensities.
- Paired comparisons on very similar samples.
- Referring back to reference(s).
- Feedback on performance.

5.3.2.5 Performance check

Ideally, when time and budget permit, a performance check should be carried out to confirm that the panel is performing in a consistent and reliable way prior to undertaking any studies. This can save money and time in the long run, particularly for large and complex studies, as issues can be identified and rectified early in the process. The full assessment protocol, including the final ballot and sample serving protocol, often using a representative subset of samples, is carried out and statistical treatments applied to check on assessor performance (see Section 5.3.2.7). Feedback, additional training and further performance checks are carried out as required.

Another way to handle poor performance, particularly when time is a factor, is to remove poorly performing individual assessors at the data analysis stage. Analysis will also identify attributes that are not discriminating between samples so that further training can be undertaken on these attributes or they can be removed from the study.

5.3.2.6 Data generation

Samples in the study are assessed by the panel using the assessment protocol. Samples need to be prepared according to strict protocols (see Section 4.4) and presented to assessors according to the experimental design (see Section 3.7).

Prior to assessing the experimental samples, a control sample may be assessed to remind and calibrate assessors to attribute qualities and intensity ratings.

The experimental design may allow for a blind control sample(s) to be included that will enable sessions and studies to be compared and/or combined. It is normal practice to carry out replicate assessments within a study; duplicate or triplicate assessments are most common.

102 Sensory evaluation

5.3.2.7 Data analysis and reporting

Data analysis is carried out to check the quality of the data produced and to assess the differences between samples. Results of the study are then interpreted and reported (see Section 6.1).

5.3.2.7.1 Checking data quality: panel performance

Consistency within and between assessors can be used to determine data quality.

Accuracy

This is the measure of how close the assessor or mean panel data are to the 'true' value. In descriptive studies, a 'true' value can often be provided only by a standard/reference sample or spiked sample. Values can be compared directly or displayed graphically using a line graph or bar chart of intensity score on the y-axis vs. assessor on the x-axis.

The value or cut-off for what is considered a 'good' or 'reasonable' result will depend on the homogeneity of the samples and may also be affected by the assessors' level of experience, their ability to understand and perceive the attribute, their use of the rating scale and the overall difference in samples. Differences in rating of around 10–20 scale units (on a 0–100 scale) away from the 'expected' result are commonly considered acceptable. Acceptable tolerances are generally given as scale units rather than percentages of the expected figure, as the latter translate into very tight tolerances at low intensity ratings, e.g. 10% of 90 is ±9 units whereas 10% of 10 is ±1 unit.

Another way to consider the accuracy of the data is to compare individual assessor's results to one another and the panel mean. A very simple way to assess the variation in the data set is by calculating the *coefficient of variation* (CV), which describes how far the data points are from one another.

$$CV = (standard\ deviation/mean) \times 100$$

To investigate panel consistency, the CV can be calculated for the range of assessor mean scores for each sample. This would represent the level of agreement between panellists for each sample. As stated earlier, the value or cut-off for what is considered a 'good' or 'reasonable' CV will depend on several factors; however, 20–30% is commonly used in sensory projects.

One-factor ANOVA can be used to investigate the difference between assessors when rating the same sample (see Appendix 5 and Appendix 10). In this instance, the one factor would be assessors, where one sample is rated on multiple occasions by each assessor. The ANOVA compares the performance of assessors (variation between assessor means) to variation within the assessors. Typically, a significance level of 5% is applied to the test and a p-value of less than 0.05 would confirm that there is a significant difference between assessors. Ideally, there should be no significant difference between assessors; however, in practice this is fairly common due to different use of scale.

One-factor ANOVA is time-consuming as it must be completed for each assessor–attribute combination; however, some software packages do include this output. More typical is to assess the assessor's use of scale from two-factor ANOVA applied to the complete data set (see Appendix 5 and Appendix 10).

Reliability (validity)
An important aspect of reliability is the ability of the individual assessor to discriminate between products. In common with 'accuracy', this is best assessed using one-factor ANOVA for each attribute. In this instance, it is applied to a data set of replicate judgements, from one assessor, for all samples (the 'one factor' being the samples). These replicate judgements are used to investigate variation between samples, i.e. has the assessor discriminated between them. Typically, a significance level of 5% is applied to the test and a p-value of <0.05 confirms that the assessor is discriminating.

The results for each assessor must be considered and compared to the panel results. It may be that none of the assessors can discriminate between the products for one or more attributes as they are very similar. Another possibility is that one or two assessors may be less discriminating than the rest of the group; however, this may not affect the discrimination from the overall panel result. The worst scenario is that nondiscriminating assessors affect the results from the panel. Where this happens, the cause must be investigated and a decision made on how to proceed (see 'Handling problem data').

Interaction plots and scatter plots are another common way in which the reliability of assessors is judged. Both of these represent graphical illustrations of assessor's ratings (individual replicates or more commonly mean ratings) for each sample. All assessors' data are represented on one $x-y$ plot which allows direct comparison of their results (see Figure 5.13); of particular importance is the relative order of their intensity ratings (rankings). This plot is further supported by two-factor ANOVA on data with experimental replicates as this allows the calculation of variation due to assessor–product interaction and the determination of its significance. A significant interaction ($p < 0.05$) is indicative of problems with assessors, samples or attribute understanding and must be investigated further (see 'Handling problem data').

Precision
This is a measure of reproducibility, or consistency, in assessor's replicates and/or panel mean replicates.

A simple way of judging precision is to calculate standard deviation or standard error as a measure of data variation (dispersion) across replicates for each assessor, sample and attribute combination (see Section 3.8.5.2). The more reproducible the results are, the smaller the value for standard deviation/error. This information can be displayed graphically as a bar chart (see Figure 5.10) showing results for all judges, including deviation/error bars. The bar chart will show which judges are the least/most reproducible, which samples cause the most problems with reproducibility and which attributes cause the most problems with reproducibility. Alternatively, CV can be calculated for replicate assessments from each assessor.

As stated previously, for assessing validity, one-factor ANOVA can be applied to a data set of replicate judgements from one assessor for all samples. This calculates variation due to one factor (in this case samples);

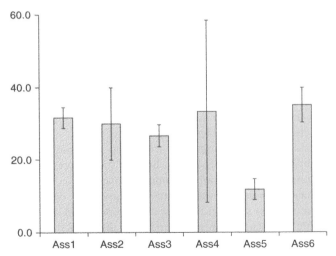

Figure 5.10 Bar chart showing mean values for each judge when rating sample A for sweetness; error bars show the spread of data from three replicate judgements for each assessor.

Sensory test methods 105

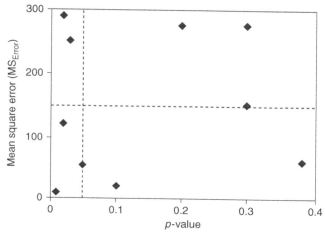

Figure 5.11 The figure shows the $x-y$ plot of p-value vs. MS_{Error}. High values for p show poor discrimination between the products ($p > 0.05$); high values for MS_{Error} show poor reproducibility across replicates.

all other variation (between replicates) is considered as background noise (error). Mean square error (MS_{Error}) can, therefore, be used to assess the reproducibility; higher values of MS_{Error} show poor reproducibility. It is common for the p-value and MS_{Error} to be plotted on an $x-y$ plot for each attribute (see Figure 5.11). The plot is generally divided into four quadrants; assessors appearing in the top right-hand quadrant relating to high p-value (poor discrimination) and high MS_{Error} (poor reproducibility) are the worst performers.

5.3.2.7.2 Handling problem data

Having identified problems with accuracy, precision or reliability, it is important to develop a consistent strategy for dealing with them. First, the cause of the problem should be identified – it is not wise to assume that always the assessor is at fault. The following are some commonly asked questions:

- Did the assessor receive correctly labelled samples in the correct order?
- Was the experimental protocol followed?
- Were there any problems reported that day that could explain the problem?
- Were there any problems with sample homogeneity?

- Is the assessor experiencing any medical or personal problems that could explain their performance?
- Does the assessor understand and perceive the attribute?

Whenever possible, assessors who provide data outside of the acceptable tolerance should be retrained. If the problems occur during training or panel validation, then this is easily achieved. If the problems, however, arise in a final data set that cannot be repeated and further training is not possible, then a decision needs to be made about the possibility of removing the assessor from the data set.

Generally, it is considered unacceptable to remove data without a compelling reason. For example, if an assessor reported having health or personal problems after the assessment, then their data can be justifiably removed. If assessors are shown to be extreme outliers and their results are affecting the overall conclusion then it is acceptable to remove their data. It is vital, however, that this is documented in all paperwork and, ideally, the analysis performed on both the full and the modified data sets.

5.3.2.7.3 Determining differences between samples

Statistical differences between samples are determined using parametric statistics. Typically, ANOVA and MCTs are used to determine which factors (samples, assessors and other design factors) cause significant variations in attribute means (see Appendix 5 and Appendix 10).

5.3.2.7.4 Displaying sensory data

It is useful to display sensory descriptive data in order to simplify interpretation, illustrate results and help communicate findings.

Sensory profiles

Spider plots (radar plots, star charts) are the traditional methods used to display sensory profiles, enabling them to be viewed and compared (see Figure 5.12). The plot is in the form of a spoked wheel. Each attribute is represented as a spoke on the wheel, with attributes placed in a logical order. The centre of the wheel is zero perceived intensity, with intensity increasing towards the circumference. Attribute means for each sample are plotted on the spokes and joined with a continuous line. Standard deviations or confidence intervals may also be plotted in a similar fashion. Several sample profiles may be overlaid for comparison and significant differences can be indicated.

Sensory traces are another way of representing and comparing sensory profiles. Attributes are marked along the x-axis, typically in the order in which they are perceived. The y-axis represents perceived intensity and

Sensory test methods 107

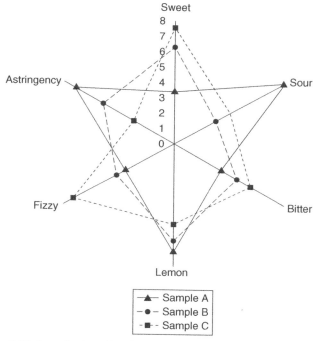

Figure 5.12 Example of a spider plot.

attribute means are plotted on it. Means are joined using a continuous line to produce a 'trace'. Similar to spider plots, standard deviations and confidence intervals can be included and several traces may be overlaid.

Interaction plots (see Figure 5.13) can be used to investigate assessor performance by plotting attribute ratings against samples for individual assessors.

Relationships between variables
The relationship between two variables is illustrated using x-y plots. In sensory analysis, perceived intensity is often the dependent variable plotted on the y-axis against independent variables on the x-axis, such as physicochemical parameters (e.g. concentration, temperature, and so on) or process parameters (see Figure 5.14). Scatter plots of data points are useful for looking at trends. Curves may be fitted to the data using regression analysis and included on the plot, so that the relationship between variables can be visualised.

108 Sensory evaluation

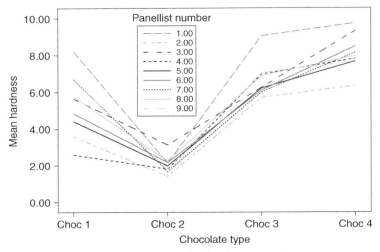

Figure 5.13 Example of an interaction plot showing reasonable assessor agreement.

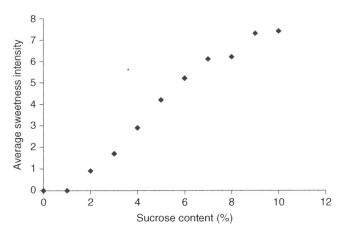

Figure 5.14 Example of $x-y$ plot.

PCA is a multidimensional statistical technique used for describing relationships between multiple variables. Large amounts of data can be simplified by identifying patterns of correlated variables and reducing them into a smaller number of underlying variables, called principal components. The output is in the form of a multidimensional map. It is

Sensory test methods 109

beyond the scope of this book to describe in detail how PCA is performed, but it is commonly available in many statistical computer programs.

PCA is particularly useful for analysing data from sensory descriptive analysis, as the data consist of a large number of attribute variables. An example of a sensory PCA map (biplot) is shown in Figure A10.4b. The map displays samples as points, and attributes as vectors. Those that are spatially close are more similar. The text in Figure A10.4b illustrates how to interpret a PCA map. PCA maps are especially beneficial when applied in market overviews, where they enable the sensory properties of numerous products in a market category to be compared in one pictorial representation. They are also helpful in determining how physicochemical properties relate to sensory properties.

5.3.3 Practical issues in dealing with long-standing panels

All panels will need some refresher training (maintenance) at regular intervals to remind assessors of attributes and intensity references and to prevent drift of ratings over time. If the panels are used intermittently, for example, with breaks of several months between studies, they will need refresher training prior to starting a new study.

After a period of time, sensory assessors begin to build a frame of reference. They tend to build up an attribute list across the course of several studies and select attributes from that list, rather than generating attributes from scratch for each new study. This can bring time advantages but care must be taken that attributes from previous studies/products are applicable to new studies/products.

There is also a tendency for assessors to move towards a fixed intensity scale, particularly if they are working within a limited product repertoire, so that although it may appear that assessors are recalibrated for each study, in reality they are not. Dummy or spiked samples can be used to check that the scale is being used appropriately.

Care must be taken when assessors are moved from one panel to another to ensure they are given appropriate retraining. In some cases, this will not be possible. For example, it is not recommended to use an assessor who has previously worked on Spectrum™ methodology for other types of scaling, as the frame of reference is firmly established and difficult to change.

When the task is repetitive, assessors may remember assessments and produce 'remembered' ratings rather than real assessments. Long-standing panels can often become demotivated by the repetitive nature of the task. The introduction of variety can be used to overcome these situations. Motivational methods are covered in Section 4.5.5.

110 **Sensory evaluation**

5.3.4 Types of descriptive methodology

A brief description of some methodologies (see ASTM MNL 13 and standard reference books for overviews), together with a case study for Quantitative Descriptive Analysis (Appendix 10), is given in the following sections. It is important that the method used meets objectives, is appropriate for the number of samples and the analysis of data required. Many sensory laboratories use modified versions that are tailor-made to suit their specific requirements.

5.3.4.1 Consensus profiling

Assessors work as a group to agree on attributes and intensity ratings. It is relatively quick, but may be subject to bias, as assessors with stronger personalities may dominate the agreement process. No statistical analysis can be carried out.

5.3.4.2 Flavor Profiling®

Flavor Profiling® was developed by Arthur D. Little Co. in 1949 (Cairncross and Sjöstrom 1950). Aroma, flavour and mouth-feel are assessed in terms of quality; intensity on a 5-point absolute 'degree of intensity' category scale (0,)((threshold), 1, 2, 3); order of appearance; aftertaste and overall impression. A panel of 4–6 selected and trained assessors assess samples individually and then discuss their evaluations as a group to determine a consensus score. The method does not lend itself to statistical analysis. Profile Attribute Analysis® (PAA®) (Neilson et al. 1988) was a later modification to the method that included the assessment of visual, tactile and auditory attributes and the use of individual assessments, category/line scales and data analysis using ANOVA.

5.3.4.3 Texture Profiling®

Texture profiling® was developed by General Foods in the early 1960s (Brandt et al. 1963; Szczesniak 1963; Szczesniak et al. 1963) and modified several times. Texture and mouth-feel properties of foods, including mechanical, geometric, fat-related and moisture content–related attributes, are assessed in terms of (i) quality – using a predetermined list of attributes, i.e. hardness, fracturability, chewiness, gumminess, adhesiveness, viscosity and geometric structure; (ii) intensity – originally on the same scale as Flavor Profiling and later on a 13-point universal scale with references for each point on each attribute scale and (iii) order of appearance from first bite to complete mastication. Originally, a panel of 6–10 assessors were selected for their ability to discriminate textural differences on the product type to be assessed. They were then trained using the predetermined scales described earlier, and worked in consensus which meant that data could not be statistically treated. In a later modification, however, they worked individually using other types of scales. The methodology was also applied to nonfood categories.

5.3.4.4 Quantitative Descriptive Analysis®

QDA® was developed at the Stanford Research Institute by Stone and Sidel (Stone *et al.* 1974) to provide descriptive data that could be analysed statistically, in contrast to the methods described earlier. It can produce a full qualitative and quantitative sensory description. Assessors (8–15), selected for their ability to describe and discriminate products in the category to be studied, agree on a list of qualitative attributes and then work individually to rate the attributes on a line scale with indented anchors. Assessors receive limited training and the primary aim for assessors is to be consistent within themselves rather than with the rest of the panel. It is, therefore, a relative assessment method. The panel leader facilitates discussions rather than leads them. Assessments are made in replicates of 2–6 repeat evaluations, data are translated into mean scores and stat-istically analysed using ANOVA, individual assessor performance is monitored and compared to that of the panel, and results are presented graphically in spider plots. The attributes are said to be closer to the language a consumer might use. QDA® is a versatile technique that can be used across a range of applications. Many sensory laboratories use modified versions of QDA® that involve more training and calibration of assessors on quality attributes and intensity rating, which may lead to fewer replicates (see case study in Appendix 10).

5.3.4.5 Spectrum™ method

Spectrum™ methodology (Meilgaard at al. 2007) was developed by Civille and takes many elements from Flavor Profiling and Texture Profiling. A full qualitative and quantitative description can be produced. Sensory qualities are assessed using a predefined, standardised lexicon from which terms are selected. The lexicon includes technical terms and terms that are applied across all products. Perceived intensity is assessed with a 15-point numbered absolute scale that can be universal or product specific and is anchored at multiple points with well-defined references, some of which may be branded products. Selected assessors (12–15) receive in-depth training on attributes and intensity references so that they all make assessments and score similarly. The panel leader plays an important leadership role. Assessors agree on the attributes and order of assessment and then rate intensity individually. ANOVA is used for data analysis. Spectrum™ method can be used across a range of applications and is particularly useful when data from different studies need to be compared. This method is more popular in the United States, partly because the references are based on American products and brands, which can be difficult to translate to equivalent products/brands in other countries. The lexicon is said to include more technical terms. It requires a high degree of panel training and maintenance to achieve the necessary level of interassessor conformance. Absolute calibration is not possible for some attributes due to differences in individual perception, e.g. bitterness and musk odours.

5.3.4.6 Other methods
Free choice profiling
Free choice profiling was developed in the United Kingdom in the 1980s (Williams and Langron 1984) and can produce a full qualitative and quantitative description. This method uses untrained assessors and can also be run with consumers. A modified repertory grid technique is used to generate individual attribute lists for each assessor. Each attribute is then rated for intensity by that assessor (the total number of attributes rated by each assessor may vary). No consensus or rationalising of terms is required and, therefore, the process of generating data is considerably faster compared to other techniques. Data analysis, however, requires the sophisticated technique of generalised procrustes analysis (GPA), which groups similar terminology and adjusts for individual scale use to create a consensus space, in which each individual's data are plotted. Results can be difficult to interpret as consumers are idiosyncratic and use inconsistent language.

Flash profiling
Flash profiling (Dairou and Sieffermann 2002) was developed as a quick sensory profiling method for industry. It is a descriptive method derived from free choice profiling in which each subject chooses and then uses his/her own words to comparatively evaluate a product set. Assessors who are experienced in sensory evaluation are generally recruited to participate as they have the necessary ability to articulate their perceptions and understand the methods used. They require very little further training as they use their own terminology and are only required to rank products for each attribute. Assessors rank all products attribute by attribute, with ties allowed, and the data are then analysed using GPA.

Quantitative flavour profiling

Quantitative flavour profiling (QFP) was developed by Givaudan (Stampanoni 1994). Flavour characteristics are assessed using a predefined lexicon for different product categories, from which terms are selected for each study. Intensity is assessed using a line scale, with indented marks at either end, labelled weak and strong. End-of-scale intensity references are used for each study. ANOVA is used for analysis.

Difference from control profiling (Deviation from reference profiling)

Assessors rate the degree of difference of a test sample from a reference sample on a range of attributes using a degree-of-difference scale (Larson-Powers and Pangborn 1978). This technique is useful for QA/QC work.

Intensity variation descriptive method

This method was developed by Gordin (1987) (as reported in Lawless and Heymann 1998) to provide information on changes in attribute intensities during consumption of products for which speed of consumption is variable by individual, e.g. cigarettes. Assessors evaluate products at specified locations in the product rather than at specified time intervals using standard descriptive methodology.

5.3.4.7 Time intensity methods

TI techniques are used to measure changes in sensory perception over time. They give additional information over traditional descriptive techniques (described earlier), such as length of sensation, changes in quality of sensation and differing intensities in quality over time. They are particularly useful for analysing the following:
- Products with a longer-lasting or changing sensory experience, e.g. chewing gum and fragrance.
- Products that themselves change over time through use or otherwise, e.g. changes such as melting or drying, changes in texture during chewing and development of wine flavour on exposure to oxygen.
- Changes in perception caused by changes in the sensory system over time, such as adaptation, e.g. sipping coffee over the course of a cup, eating a spicy dish and smelling an air freshener.

They can be used to measure the following:
- A single sensation.
- Multiple sensations.
- Multiple exposures within a single measurement period.
- Qualitative changes in perceived sensation.

114 Sensory evaluation

- Hedonic changes in perceived sensation. Only basic training on the use of the measurement apparatus is given to assessors in this case.

Figure 5.15 illustrates a typical time intensity curve and shows some of the sensory parameters that can be measured using this technique. Definitions of these parameters are as follows, based largely on ASTM standard E1909-97(2003):

- I_{max}: Maximum intensity.
- Plateau: Perception remains at maximum intensity for a length of time.
- T_{init}: Time of exposure to stimulus.
- T_{onset}: Time of start of perception.
- T_{max}: Time from exposure to stimulus (T_{init}) to maximum intensity.
- T_{plat}: Time that perception remains at maximum intensity.
- T_{ext}: Time from exposure to stimulus (T_{init}) to time to end of perception.
- T_{dur}: Duration of perception. Time from onset of perception (T_{onset}) to end of perception (T_{ext}).
- Lag time: Time from introduction of stimulus to perception (T_{init} to T_{onset}).
- Rate of increase in perception: Rate of increase in intensity from time at start of perception (T_{onset}) to the start of the plateau. Can be derived using I_{max}/T_{max}.
- Rate of decrease in perception: Rate of decrease in intensity from time from start of decreasing perception to end of perception (T_{ext}). If there is no plateau this can be derived as follows: $I_{max}/(T_{dur} - T_{max})$.

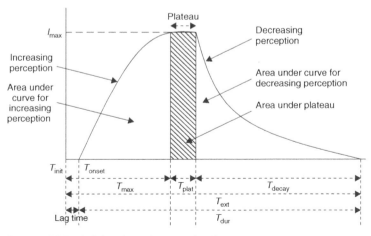

Figure 5.15 Typical time–intensity curve showing parameters.

Sensory test methods 115

- T_{decay}: The time from maximum perceived intensity (I_{max}) to end of perception (T_{ext}).
- Area under curve.
- Area under curve for increasing perception.
- Area under curve for decreasing perception.
- Area under plateau.

When designing a TI study consideration needs to be given to the following:

- Which TI parameters are of interest. For example, time to sweetness onset is important for products containing artificial sweeteners; time to extinction is important for chewing gums.
- The usefulness of TI over alternative methods which may be less time-consuming and costly, e.g. a single measure of persistence in QDA may provide sufficient information without the need for TI.
- The length of the measurement period, e.g. to extinction of sensation, to a specified time and to a specified intensity. It should be sufficient to cover key sensory changes.
- Whether to make measurements over one ingestion/use or over the course of multiple ingestions/uses.
- Which attributes to include ensuring that key sensory changes are covered.
- Which elements of the procedure to standardise, e.g. standard chew rate, standard rubbing time intervals for topical applications and set time for swallowing.
- Appropriate measurement intervals, e.g. if sensations do not change much over the duration of perception, the measurement intervals may be longer.
- Appropriate data recording intervals.
- Experimental design. Typical designs can be used, but the number of samples may be limited by the length of the assessment.

Statistical analysis can be complex and may involve curve averaging, which is beyond the scope of this book.

For more detailed information see Dijksterhuis et al. (1994), Lawless and Heymann (1998) and Liu and MacFie (1990).

Discrete-point time intensity

In this technique, a minimum of 10 assessors rate intensity of sensory characteristics at specified points during the sensory experience. This is performed using the typical descriptive techniques described earlier, e.g. QDA®. This can be at any of the following:

- Set time intervals, e.g. every minute cued with an instruction or sound.

116 Sensory evaluation

- Different points in the ingestion process, e.g. first bite, first chew, after complete mastication, immediately after swallowing. Assessors may be instructed to complete certain actions at a standard, cued time.
- Different phases as the product changes during use, e.g. upon opening a bottle of wine, after pouring the wine, after breathing for a set time period. This may be complex, as phases may occur at different time points, e.g. complete melting in the mouth will vary depending on individual in-mouth temperature and physiology.

This technique has the advantage that assessors trained in typical descriptive techniques can perform the test with limited additional training. Multiple attributes can be measured at one time; the number depends on the length of time between measurement points, i.e. what number is practical to assess in the time interval. Discrete point TI is also useful when sensations change slowly over time, or, where a sensation is measured for a longer period of time, e.g. in excess of 5–10 minutes. In this instance, assessors become less reliable at continually measuring perception (continuous TI techniques – see next section), as the frame of reference becomes less well remembered; this leads to increased variability. It is easier and less costly than the TI techniques described here.

Data can be analysed at each time point using the data analysis techniques described earlier for descriptive analysis, i.e. ANOVA. In addition, curve averaging techniques can be used to plot intensity over time.

Continuous time intensity

In this technique, assessors continuously rate intensity of sensory characteristics. A minimum of 10 assessors is recommended; however, other texts make reference to as few as six assessors. A continuous method of recording data is used, such as a pen on a chart recorder or, more typically, a computer with a mouse or joy stick.

Specially trained assessors are essential. It is acceptable, and can save time, to use assessors who have been trained previously to carry out descriptive analysis, e.g. QDA®. Training steps for continuous TI include familiarisation with the method, equipment and procedures; familiarisation with different time–intensity profiles using examples; learning to focus on a single attribute, so that it can be identified and quantified in a complex system; accurately recording changes as they occur and carrying out multiple tasks at the same time, e.g. chewing and recording.

Quality and intensity references may be used. These must be remembered, as it is difficult to introduce them during the assessment.

Panel training checks can include assessing variation around intensity at set time points, with emphasis on I_{max}. Individual consistency is more important than consistency with other assessors, as the technique inherently produces data that are variable between assessors. Assessors typically produce time–intensity curves with distinct individual 'signatures', thought to be due to differences in physiology and scale use, that persist even after extensive training.

Continuous TI gives more information than other descriptive techniques. Only one attribute can be measured at a time (although a little-used variation of this technique has been developed to assess two attributes at a time). This means the technique is slow and there is evidence to suggest that an attribute assessed in isolation may be assessed differently than when assessed at the same time as other attributes. It cannot be used when a very long sensation would cause excessive variation in individual data due to a high memory load. It is more complex and more costly than other descriptive techniques.

Typically, curves are averaged (see Lui and MacFie 1990) and data are analysed using ANOVA on selected time points or curve parameters at each point/parameter, with at least three factors of time, assessor and treatment. Additional factors of replicates and further design variables may also be included in the analysis.

Further data analysis requires sophisticated statistical techniques and is beyond the scope of this book. *For more detailed information (2003) Standard Guide for Time-Intensity Evaluation of Sensory Attributes.*

Temporal dominance of sensations

Temporal dominance of sensations (TDS) tracks multiple sensory attributes over time and is able to detect sequences of dominance of sensation. It is well suited to multivariate investigation and is relatively quick. This is in contrast to traditional TI techniques, which measure intensity of sensations individually and as if they were perceived independently. In TDS, 16–30 assessors are trained to recognise approximately 10 attributes, and to continuously assess which attribute is dominant and to rate its perceived intensity on a line scale. During data collection, the trained panel work independently and continuously over time, selecting the attribute that is dominant from a list of approximately 10 attributes and rating that attribute's perceived intensity. Data are collected in duplicate. Mean panel values for 'time at start of dominance', 'perceived intensity rating' and 'duration of dominance' are collated in product $\times$ attribute matrices. Data can be treated parametrically and analysed using PCA. Duration of dominance of sensation for each attribute is coded by summing the number of subjects (across replications) selecting that attribute as dominant over time. Curve smoothing and significance testing are applied using specialised software. An attribute is judged as dominant if frequency of selection is significantly higher than chance. In addition to recovering the same information as TI, TDS yields additional information about time and duration of sensations that can be used in product design.

5.3.5 Dos and don'ts

- ✔ Do carefully consider short- and long-term objectives prior to setting up a study or a panel.
- ✔ Do carry out a team evaluation of all samples prior to designing a study to give better input into the design, e.g. by assessing the sensory range, and to check if samples are as expected, e.g. to check for incorrect sample production, taints and so on.
- ✔ Do consider whether the product changes over time or multiple usages; TI methodology may be appropriate.
- ✔ Do modify existing off-the-shelf descriptive methodology to meet your objectives.
- ✔ Do consider logistical implications of sample preparation and serving – this can be complex, time-consuming, take up a lot of space and require working to tight deadlines.
- ✔ Do carry out a pilot session prior to the main study – preparation and assessments can often take longer and be more logistically complicated than expected.
- ✔ Do carry out a check on panel performance prior to the actual assessment. It will save time and money in the long term by avoiding mistakes.
- ✔ Do consider the possible effect of participation in previous studies on a panel or assessor's approach to a new project.
- ✔ Do maintain regular communication with the panel outside of training and assessments – they are not machines!
- ✖ Do not ask the panel to make hedonic or acceptability judgements – they are not representative of consumers.

5.4 Affective/consumer tests

5.4.1 Introduction

Consumer testing assesses subjective responses to a product. Using both qualitative and quantitative methodologies, researchers can gain an insight into consumer preferences, attitudes, opinions, behaviours and perceptions concerning products. Consumer testing is, therefore, a key part of the product development process and can also be used to develop sensory-related brand positioning, communication and advertising.

Following product launch, consumer testing is also vital to monitor market position and, where necessary, to find avenues for product improvements or optimisation. Such testing, however, is successful only when the data are reliable and valid. Benefits can be gained from a combination of approaches. For example, focus groups can highlight important attributes for further assessment in a quantitative survey; they can also be used to probe issues that have been highlighted from the results of quantitative research. Quantitative surveys can be useful in highlighting particular consumer subsets for participation in one-to-one interviews or focus groups. Comparing the results from tests in which products are presented with branding information, to tests in which the samples are debranded, can give information on the relative strength of the communication, i.e. how effective is the branding information.

5.4.2 General considerations

For any type of consumer test, the number and type of assessors are important considerations. For quantitative tests, large numbers of assessors, a minimum of at least 100, are required if the results are to be meaningfully extrapolated to the larger population. The following factors can be considered when selecting consumers:
- Geographical regions
- Demographics
- Psychographics
- Lifestyle/stage
- Product usage/nonusage

In general, employees should not be recruited for consumer testing as they may be biased as a consequence of product knowledge.

Test location will be driven by the research objectives, the available budget and an appreciation of the relative merits of laboratory test, CLT and HUT. Table 5.3 summarises the potential advantages and disadvantages of each location.

In addition to HUTs, there are other situations that allow products to be tested in a natural context, e.g. purchase from vending machines, restaurants, hospital bed.

Timing of consumer tests should also be given consideration. For example, avoid Christmas and summer vacation when consumer attention is focused elsewhere. It is also better, where possible, to test seasonal products at appropriate times of the year. Some products may even require consumers to be in the appropriate need state. For example, sports drinks may need to be tested during and after exercise.

Sensory evaluation

Table 5.3 Advantages and disadvantages of test locations

	Advantages	Disadvantages
CLT: Laboratory test	• Relatively high response rate • Controlled conditions • Immediate (computerised) feedback • Low cost • Several products can be assessed per consumer	• Not representative of the natural context • Important attributes can be missed • Number of questions that can be asked is limited • Respondents not always representative of population
CLT: Hall test	• High number of respondents • Respondents from general population • Several products can be assessed per consumer • More control over how product is tested	• Unrepresentative surroundings • Less control than in a laboratory test • Important attributes can be missed • Number of questions that can be asked is limited
Home use test	• Relatively high number of respondents • Product tested under real conditions • Ability to test product under repeated use conditions • Ability to gain realistic information concerning intention to purchase	• More nil returns and missing responses • No control over product use • Time-consuming • Slow feedback • Small number of products • Generally more expensive

5.4.3 Questionnaire design

Invariably, consumer research involves the completion of some form of questionnaire. A questionnaire not only enables the collection of accurate data from the respondent, but also provides structure and a consistent format for the collection of responses. The design of the questionnaire is a very important stage in the investigation. The following need to be considered.

5.4.3.1 Research objectives
These must be considered at all stages in the design process. This will determine the type of questionnaire and the location for its administration. Only questions relating to the research objectives should be posed. Do not ask for more than what is required.

5.4.3.2 Type of questionnaire

Determine whether a structured or semistructured questionnaire is required. For qualitative research, a flexible type of questionnaire may be more appropriate. Consider whether the questionnaire is for self-completion or for a face-to-face interview. Self-completion questionnaires must give very clear instructions to the respondent. When administered by a researcher, respondents can ask for clarification, or can be reminded to give only one answer, and so on. Note that researchers need to be fully briefed on how to pose questions, and what additional information to proffer to avoid introducing bias to the data. Keep the questionnaire as short as possible to meet the research objectives. Minimise the number of questions and products to avoid fatigue.

5.4.3.3 Layout

The layout should be logical and give structure for face-to-face interviews. Pay careful attention to the order of questions to ensure early responses do not influence later questions, e.g. preference questions should precede any additional questioning on specific sensory attributes. More sensitive information, e.g. alcohol consumption, use of hygiene products, age, and income, should be asked towards the end of the questionnaire, when the respondent is feeling more relaxed. Note that the use of 'response bands', e.g. age 25–40, can desensitise questions relating to income and age, etc.

5.4.3.4 Type of question

The research objectives should drive the type of questions to be included; questions tend to fall into the following three categories:
- Behavioural – Who? Where? When? How many? How often?
- Attitudinal – What do you think of? Which is best? Why do you?
- Classification – age, gender, income, etc.

Open or closed

Whether questions are open ended or closed will be determined, to some extent, by the research objectives. If in-depth analysis is required, open-ended questions may be necessary. Open-ended questions are easy to ask but are difficult to process and analyse statistically. Closed questions are quicker to ask and are easier to analyse.

Consideration should also be given to the number of different types of scales used, e.g. hedonic, just about right (JAR), intensity and Likert, as this can confuse respondents. Similar scales should be grouped together.

5.4.3.5 Wording

Make the question as short and simple as possible.
- ✔ Do not use any jargon, e.g. loss leader, short hand, e.g. pmt, or ambiguous words, e.g. usually.
- ✘ Try not to use negative or arithmetic phrases, e.g. when do you not use.
- ✔ Simplify numbers to simple scales, such as none, less than half, half, and more than half.
- ✘ Do not ask two questions in one, e.g. was the ready meal tasty and easy to cook?
- ✔ Keep questions within respondent capabilities by avoiding questions which require considerable memory or technical understanding.
- ✘ Do not allow any hypothetical questions, e.g. If you had a pet would you feed it this?
- ✔ Do provide appropriate instruction to the assessor on how to perform the assessment, such as presenting scales with information on what the scale is measuring and how the scale should be used.
- ✔ Ensure back translation of questionnaires translated to other languages to ensure meaning is retained.

5.4.3.6 Coding

Determine how the responses to each question are to be entered into spreadsheets for data analysis. For optically read questionnaires, this is integral to the design process but it should be considered at the design stage for any questionnaires in order to make data entry and analysis efficient.

5.4.3.7 Pilot

All questionnaires should be piloted, preferably with a representative sample of individuals, or at the very least, with coworkers.

5.4.3.8 Dos and don'ts

- ✔ Do carry out a pilot of the questionnaire.
- ✔ Do make sure questions are unambiguous.
- ✘ Do not ask questions that do not relate to the research objectives.
- ✔ Do ensure that researchers conducting questionnaires are thoroughly briefed.
- ✔ Do keep the questionnaire short to avoid fatigue.

For further information on questionnaire design see Brace (2004).

The following section reviews the most common qualitative and quantitative methods used for the collection of consumer sensory data and outlines the key considerations to be made by the sensory professional when performing such tests.

5.4.4 Qualitative methods

Qualitative methods enable researchers to gain a deeper insight into consumer reaction to product concepts, their attitudes, opinions and preferences towards products and often to define the critical attributes of a product from the consumer perspective. Various methods are used including one-to-one in-depth interviews, group interviews and, most commonly, focus groups. Ethnographic research techniques, where the researcher directly observes or even lives with subjects in their natural environment, are also useful ways of obtaining qualitative data concerning consumer interactions with products. Typical examples include observing purchase behaviour in supermarkets or product use behaviour, e.g. how people apply make-up or use personal care products. The widely used focus group technique is considered here. *For further information see Ereaut et al. (2002) for a series of books on qualitative techniques from the Market Research Society.*

5.4.4.1 Focus groups

Objective: Focus groups can be employed to meet various objectives. These include formulating a hypothesis, testing the feasibility of a new product concept, testing communication strategies via packaging or adverts, developing items for inclusion in questionnaires, identifying critical sensory attributes for a product category and probing issues highlighted in quantitative research.

Rational: In contrast to quantitative techniques, a more in-depth analysis can be obtained. Focus groups can provide an insight into the reasoning behind consumer perceptions and decisions through probing of their initial responses and observations of their actions.

Experimental design: Generally 8–12 participants are recruited per group. Dependent on the objective, these may be representative product users, anticipated users, nonusers or even individuals holding certain attitudes or beliefs, etc. Consideration should be given to the homogeneity of the groups, e.g. if regional differences are expected, focus groups should be conducted in different parts of the country. At least three focus groups are recommended to enable comments concerning the consistency of the results to be made, although conflicting opinions across groups can, itself, be viewed as a result.

124 Sensory evaluation

Procedure: A trained moderator produces a discussion guide to ensure that the focus group covers all the key issues relevant to the investigation. Figure 5.16 shows an example of a discussion guide for an investigation into the use of skin creams. The potential bias of the moderator is one of the main criticisms of this technique; therefore, recruiting a trained

---

FOCUS GROUP PROTOCOL – Skin cream application
- **Arrival and welcome** – refreshments, sit in lounge (background music)
- **Ensure all volunteers have filled in consent form/confidentially agreement**
- Enter room, ask everyone to sit around table – sit in line of camera

INTRODUCTION (10 minutes): Introduce self, note ground rules – mention videoing. TURN ON DICTAPHONE AND VIDEO. (Remind everyone why they are there – to find out likes and dislikes of skin creams, explain **why** we do this type of research. ASK PEOPLE TO SPEAK CLEARLY AND EXPLAIN THAT EVERYONE'S OPINIONS ARE IMPORTANT, ALL VALUABLE – NO RIGHT OR WRONG ANSWERS. Explain about break.)
Warm up: Go round table and state name, WHERE FROM and what type of creams you buy or favourite type of cream.

RAPPORT (10–15 minutes): Discuss skin cream category. Hand cream. What's out there? What's more popular? What's changed in the past 5 years? **I want to know what attributes you like about the creams therefore........ When purchasing skin creams what do you look out for?** (added vitamins, perfumed, fragrance free?, packaging – small, handbag sized or bulk cheaper packs?) **Brainstorm on White Board**

IN DEPTH (60 minutes): Uncover creams. **TRY ON FIRST CREAM** – ensure volunteers try on as would normally, for example at home remind them to wash their hands afterwards: RELATE WHAT PEOPLE SAY BACK TO PREVIOUSLY MENTIONED 'WHAT THEY LOOK OUT FOR IN CREAMS'. Try next cream – after trying three, move on to next questions relating back to convenience, etc.
Probe issues: Convenience, costs, variations, family likes and dislikes: is the make/brand important to you?
- Raise the issue about packaging – does this play a major role in your choice of skin creams?
- What about thickness? Why are they preferred?
- Discuss pros and cons. Probe important sensory attributes. Reasons for likes and dislikes.

NB Break (after 45 minutes or 3 creams – 10 minutes – then do last 3 creams with similar questioning).

CLOSURE (10–15 minutes)
Review concept and issues. Ask for clarification.
Ask for new product suggestions or variations on the theme.
Last chance for suggestions. False close (leave room to collect payment).
Close, thanks, distribute incentives, dismissal.

Figure 5.16 Example of a discussion guide for a focus group on the use of skin cream.

moderator is key to the success of a focus group. The group must be able to manage the discussion but remain impartial, and ensure all participants contribute whilst allowing no individual to dominate.

A focus group discussion is divided into four stages: introduction, rapport, in-depth and closure. The discussion is normally recorded by audio or video tape and/or viewed in real time through a two-way mirror and/or internet/video link. Participants must always be informed that this is the case.

The *introduction* serves as warm up with the key aim of explaining the objectives of the discussion session to the participants. The moderator should explain his/her role, how the session is to be recorded and general rules about the discussion, for example, that only one individual should speak at a time and that there are no wrong answers. This typically lasts about 10 minutes.

During the *rapport* stage, typically 10–20 minutes, general issues concerning the topic should be discussed. This enables the key issues to be considered early on but also allows participants to settle into the discussion.

The *in-depth* discussion then follows, in which various techniques and stimulus materials can be employed to probe particular aspects of the discussion and draw out participant opinions, experiences, etc. This stage can last up to an hour.

Finally, in the *closure* stage, the moderator should check that all the key points have been covered/raised and then close the session. Sometimes a period of false closure is included to allow additional points to be elicited. This stage usually lasts for about 10 minutes.

The recording of the discussion is then transcribed and a report written.

Data Analysis: Quantitative analysis, such as tabulating the frequency of responses to particular questions, is not appropriate as sample sizes are so small. It is not possible to generalise the results to a larger population. It is possible to highlight general trends in responses and exemplify these with quotes. It is, however, important to pay attention to the process through which ideas were formulated as this can be more illuminating than the quotes themselves. Particular themes can be extracted, as can general concepts emerging from the data. Specific software is now available to facilitate this task. It is possible that the researcher had initial hypotheses about the type of information that would evolve from the focus group; comments can be made in comparison to these. Table 5.4 shows an extract summarising comments from two focus groups on a particular skin cream. The results from focus groups can also be supported/contrasted with secondary sources, such as marketing reports and opinion polls.

126 Sensory evaluation

Table 5.4 Extract summarising comments on a particular product from focus groups evaluating skin creams

Product 3

General

Positive

Aroma – liked, fresh, love smell

Consistency – liked, nice and thick

Effectiveness – absorbed quickly, not sticky or greasy

Packaging – good size for desk, also squeezy tube for handbag

Negative

Aroma – weird; intrusive

Individual quotes

Aroma
'Fresh smell'; 'reminds me of summer'

Packaging
'The packaging is really fresh, clean, no worries, very simple but very effective'

Colour
'I generally don't mind what colour it is as long as it doesn't stay that colour on your face'

Effectiveness
'Cooling feeling to skin – feels fresh'

Consistency
'Due to the thickness it might go crusty if left in the cupboard for a long time'

Effectiveness
'It absorbs really well and isn't at all greasy'

Summary

The majority of the group liked this cream. The fact that it is multifunctional (for face, hands and body) was a clear benefit of this product. The group also favoured the simple but effective packaging and the fresh fragrance. The rapid absorption and lack of greasiness were other benefits commented on.

Conclusions: It is important to emphasise that, due to small sample sizes, the conclusions from focus group data cannot be generalised to the larger population. However, it is possible to conclude key themes and trends, or identify a range of attributes driving liking of a product which can then be used for further quantitative consumer studies.

Dos and don'ts

✔ Do use a *trained* moderator.
✔ Do prepare a discussion guide.
✔ Do consider giving participants tasks to do before focus group sessions.
✘ Don't apply statistical analysis to the data.
✘ Don't attach too much importance to comments made by individuals.

5.4.5 Quantitative methods

Quantitative consumer testing is used to measure either preference or acceptance of products. Preference implies some form of hierarchy but does not necessarily imply that the consumer likes the product, whereas acceptance testing gives an indication of the magnitude of the level of liking of the product. Diagnostic testing is used to understand consumer preference and acceptability.

5.4.5.1 Preference tests

Paired comparison and ranking tests are techniques used to determine if differences exist between two or more products for a particular attribute and are outlined in Sections 5.2.3.1 and 5.2.3.3, respectively. If the attribute tested is 'preference', then these techniques can be successfully used to gain insight into consumer preference.

Objective: To determine if a significant difference exists in preference between two (paired preference test) or more than two (ranking test) products.

Rational: Preference tests provide evidence of whether one product is preferred over another. This can be useful when looking to verify an improved formulation or measure performance against competitors. It is difficult to make an exact match, and positive preference for a modified product may be a more appropriate objective than determining that no significant difference exists.

Experimental design: The experimental design considerations for a paired preference and ranking test are the same as for the paired comparison and ranking test that are outlined in Sections 5.2.3.1 and 5.2.3.3, respectively. An important exception is that a much larger number of assessors, between 50 and 100, are required if valid conclusions are to be drawn. Including a 'no preference' option or 'allowing ties' is sometimes considered, as, although assessors may be able to distinguish between the products, they may genuinely have no preference. Nevertheless, the forced choice approach is still recommended as it retains more statistical power. Furthermore, many consumers are still likely to use the no preference option as a means of avoiding 'giving the wrong answer' or, an easier option than making a choice, despite the fact that they do have a preference. It is inappropriate to precede preference tests with questions concerning product attributes as this may focus assessor attention on particular attributes and, subsequently, bias their overall opinion. Additional elements can be added to assess the magnitude of preference, such as a 'degree of preference scale' or sureness judgements and R index analysis (see Appendix 11).

Procedure: The procedures for paired preference and ranking tests are the same as indicated in Sections 5.2.3.1 and 5.2.3.3, respectively. The question posed is, however, slightly different and examples of typical questionnaires are given in Figure 5.17.

Paired preference test for digestive biscuits

Assessor_____   Date_____

Please taste the two biscuits in the following order. Use the water provided to cleanse your palate before tasting each sample:

                268   921

Which biscuit do you prefer?_____ (state the code number here) You must make a choice.

Please comment on why you preferred this sample:

Thank you for your participation.

Preference ranking test for digestive biscuits

Assessor_____   Date_____

Please taste the four biscuits in the following order. Use the water provided to cleanse your palate before tasting each sample:

           348   268   921   551

Place the code numbers in the appropriate position below. One code only per line – no ties are allowed).

Most preferred _____
                _____
                _____
Least preferred _____

Comments:

Thank you for your participation.

Figure 5.17 Typical questionnaires for the paired preference test (top) and the preference ranking test (bottom).

Sensory test methods 129

Data Analysis: Data from paired preference and preference ranking tests should be analysed using the techniques outlined in Sections 5.2.3.1 and 5.2.3.3, respectively.

If a 'no preference' option has been allowed in a paired preference test, there are three possible approaches to the data analysis; the third option is rarely used.

- Ignore the 'no preference' responses. This will reduce the number of assessors and, consequently, reduce the power of the test. The number of no preference responses should be reported.
- Make the assumption that respondents would choose the 'no preference' option randomly and split the 'no preference responses' equally between the products. The number of 'no preference' responses should be reported.
- Distribute the 'no preference' responses proportionally according to the preference for each product, as determined from the data collected. Essentially, this implies that a forced choice approach should have been implemented in the first place.

Conclusions: The analysis of data from these types of tests enables the researcher to conclude whether a significant preference exists for a particular product, or in the rank order of the products. However, the magnitude of the difference in preference is not indicated, unless additional questions have been included. Furthermore, care should be taken when reporting no significant difference in preference testing. This does not mean that the samples are 'not significantly different', although this is an erroneous interpretation commonly made by nonspecialists. The samples may still be different; the consumer may simply have no preference for either, indeed they may dislike both.

Dos and don'ts

✔ Do include instructions that are easy to follow.
✖ Do not precede preference questions with attribute diagnostics (see Section 5.4.5.3).
✖ Do not assume that products are similar if no significant preference exists.

5.4.5.2 Acceptance tests

These tests provide an indication of the magnitude of acceptability of products. The most popular method is hedonic rating.

Hedonic rating

Objective: To determine the level of liking of one or more products. For example, ascertaining how much consumers like a new product concept, or comparing the level of liking of a standard product to the market leader.

Rationale: Preference tests give no indication of how much a product is liked. Based on the assumption that consumers will only buy a product if they enjoy eating it, asking consumers to rate a product for liking provides this valuable information.

Experimental design: Typically, 100 consumers are recruited. They are generally representative of the target market or current users. Samples are presented to each consumer, either monadically, sequentially monadically or simultaneously. As individuals are prone to scoring initial samples abnormally high, it is good practice to present assessors with a 'dummy' sample to remove this source of bias. The dummy sample should be similar to the sample set; however, its data are discarded. The remaining samples are then presented to each assessor according to a balanced, or at least randomised, design.

Procedure: For each product, subjects are asked to indicate their level of liking on a hedonic scale. A hedonic scale includes a series of verbal statements that convey a level of like or dislike. The most common is the 9-point hedonic scale designed by Peryam and Girardot in 1952. Smiley faces with more child-friendly terminology, or just pictures of facial expressions, e.g. the Snoopy scale (Moskowitz 1985), are common approaches with children. Examples of these scales can be seen in a typical hedonic questionnaire in Figures 5.18 and 5.19. Prior to further data analysis, the responses are converted to numeric values according to the number of categories on the scale. Categorical scales provide a small number of response options and, as such, may limit discrimination between samples. They can, however, be susceptible to central tendency error, i.e. assessors avoid the use of end points, further limiting the number of categories. Furthermore, although it is assumed that the intervals between the categories are equal, this is not necessarily the case and violates assumptions for the use of parametric statistical tests. The labelled affective magnitude (LAM) scale (Schutz and Cardello 2001), shown in Figure 5.20, alleviates some of these issues and provides an alternative to a categorical hedonic scale. The positioning of the labels on the LAM scale was determined through previous research in which subjects were asked to assign scores, on a ratio scale, representing the magnitude associated with each of the terms.

Data analysis: Generally, researchers determine average hedonic scores for each product and then determine whether significant differences exist between products. Typically, mean values are calculated and ANOVA is

Sensory test methods 131

Hedonic rating test for digestive biscuits

Assessor_____   Date_____

Please taste the biscuits in the following order. Use the water provided to cleanse your palate before tasting each sample:

348   268   551

Indicate how much you like the sample by ticking the most appropriate phrase below:

348

___ Like extremely
___ Like very much
___ Like moderately
___ Like slightly
___ Neither like nor dislike
___ Dislike slightly
___ Dislike moderately
___ Dislike very much
___ Dislike extremely

Comments:

Thank you for your participation.

Figure 5.18 Typical questionnaire for a hedonic rating test.

Figure 5.19 Example of a 7-point facial expression hedonic scale for use with children.

applied to the data set (see Appendix 5). However, there is now considerable evidence that the intervals between the categories on the hedonic scale are not equal and, hence, nonparametric statistics, such as calculating median and mode values and applying Friedman ANOVA (see Sections 3.8.5.1 and 5.2.3.3, respectively), should be employed. As the LAM scale has ratio properties, mean product values can be calculated and ANOVA applied to the data. Note that taking an overall average can sometimes hide subgroups within a population.

132 Sensory evaluation

Conclusions: The analysis of data from hedonic rating tests enables the researcher to make conclusions about the level of liking of a product, or make comparisons between the scores assigned to several products, in the context of the particular scale used. The score (action standard) on a liking scale, for a product worthy of consideration for the retail market, will be highly dependent on the product category. Building a database of past scores can give an insight into typical scores for products and may also highlight cross-cultural differences where international panels are used. Cultural background is known to influence the use of liking scales.

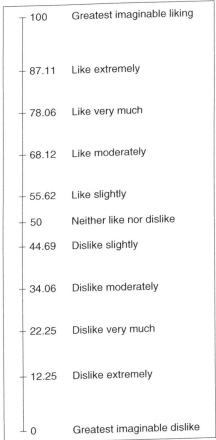

Figure 5.20 LAM scale and associated label locations. (*Note:* Numeric label locations are for information only and do not appear on the scales given to assessors.)

Remember that just because consumers like a product, it does not necessarily mean that they will buy it. Additional data are needed to make such inferences. Purchase intent scales or more novel approaches such as vending machine studies can provide this data. Furthermore, as indicated in the previous section on preference testing, the fact that samples receive similar liking scores does not mean that they are similar in terms of their sensory properties.

Example: A company was asked by a retail client to develop a biscuit product for an own brand label which will attract significantly higher scores for liking than the leading brand at a significance level of 5%. The client also wanted to know the relative position of one other own brand product in the category.

A consultancy firm was employed to recruit 120 consumers to carry out hedonic testing (9-point category scale) on three samples of biscuit at a CLT. The recruitment criteria specified that consumers should be regular users of the product, and shop at the client's retail outlet on a regular basis.

Median and mode values for each of the biscuits were determined and are given in the following table. They indicate that both the retailer's own brand and the brand leader received average scores of 'like moderately', although the modal response for the brand leader was 'like very much'. Furthermore, rank sum totals were ascertained for use in the Friedman analysis.

	Median	Mode	Rank sum
Retailer's own brand	7	7	266.4
Brand leader	7	8	279.6
Competitor's own brand	6	6	174

The competitor's own brand performed less well with average scores equivalent to 'like slightly'. To determine if the differences were statistically significant, the data were subjected to a Freidman analysis in a statistical software package which returned an F-value of 63.46 and an associated probability value of less than 0.001. The products were divided into two subsets, A, the retailer's own brand and the brand leader, and B, the competitor's own brand. It can be concluded that there is no significant difference in liking between the brand leader and the new own label brand. The competitor's own label product does, however, have a significantly lower overall liking score ($p < 0.05$).

134 Sensory evaluation

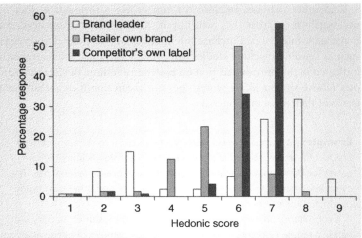

Figure 5.21 Frequency of hedonic response categories to three products.

Further information, however, can be gained by looking at the distribution frequencies of each sample as indicated in Figure 5.21. It is evident that the brand leader has more responses in the 'like very much' (8) and 'like extremely' categories (9) but, due to a subgroup of consumers who do not appear to like the product (scores of 2 and 3) both products have similar average scores. Most consumers like the retailer's own brand product but there are no scores in the top two liking categories. To attract consumers who prefer the brand leader, more development may be required. Assessing data at this level of detail provides additional information to help inform decision making.

Dos and don'ts
- ✔ Do recruit an appropriate consumer sample.
- ✔ Do use a dummy sample in the first position when comparing several products.
- ✔ Do consider the use of nonparametric statistic for data analysis.
- ✘ Do not precede hedonic rating with attribute rating questions.

5.4.5.3 Attribute diagnostics

Collecting additional data regarding consumers' perception of a product's sensory attributes can help elucidate why consumers like or dislike

Sweetness					
	Too sweet		Just about right		Not sweet enough
	☐	☐	☐	☐	☐
Graininess					
	Too smooth		Just about right		Too grainy
	☐	☐	☐	☐	☐

Figure 5.22 Examples of JAR scales.

a product. Paired preference and acceptance tests can be focused on particular attributes by adapting the question. For example, 'which of these two samples do you prefer in terms of appearance?', 'How much do you like the mouth-feel of this product?' JAR scales, attribute intensity scales, 'tick all that apply' and 'agreement' scales are also common tools used to gather attribute diagnostic data. There are some criticisms of the use of these tools. In directing assessors to particular attributes, some attributes may be missed, and those included may not be important, or even perceived in the product. It is also argued that including attribute diagnostics, particularly JARs, can affect assessor response to questions of overall liking or preference, by directing assessor attention to specific attributes. For this reason, it is highly recommended that questions regarding overall liking or preference are completed on all products, prior to any attribute diagnostics.

JAR scales: JAR scales are bipolar and typically five categories wide (Figure 5.22). Their main disadvantage is that the perception of what is 'just about right' varies across individuals.

Attribute intensity: Consumers assess the intensity of attributes, typically, on a category scale (Figure 5.23). Intensity can be correlated with liking data to see which attributes are related to liking. A possible disadvantage of this type of scale is that consumers may not understand the attributes.

Agreement scales: Assessors are given a statement relating to an attribute and asked to rate their levels of agreement on a bipolar, category 'agree' to 'disagree' scale (Likert scale).

Data analysis of attribute diagnostics: The ability of consumers to correctly use these types of scales is sometimes debated. Unlike trained assessors, different consumers are likely to use the attributes differently, use different parts of the scale and are subject to central tendency error (see Section 2.2.1).

```
Sweetness:
    □      □      □      □      □      □      □
   Not                                        Very
   sweet                                      sweet

Graininess:
    □      □      □      □      □      □      □
   Not                                        Very
   grainy                                     grainy
```

Figure 5.23 Examples of attribute intensity scales used with consumers.

JAR data are rarely normally distributed; bimodal distributions are very common. For this reason, nonparametric statistics should be used to analyse this type of data, although in practice, parametric statistics are commonly applied. The median and mode should be used to calculate averages; the arithmetic mean can give a false impression, particularly if the distribution of the data is bimodal. Analyses such as the Friedman test (see Section 5.2.3.3) should be used to identify significant differences between samples in order to obtain valid and reliable conclusions.

5.5 Linking consumer, sensory and product data

A range of techniques have evolved that enable the sensory professional to combine data from sensory panels, data collected from consumers and data related to the product such as chemical, physical, formulation and/or process variables. Such techniques provide valuable insights into the way in which sensory properties drive consumer preference and food choice behaviour, and how the product can be designed to give the sensory properties desired by the consumer. Careful experimental design and advanced statistical analyses are a key part of these techniques and, consequently, the input of a statistician is advisable from the start of such investigations. A detailed examination of these techniques is beyond the scope of this book. In view of the importance of these techniques, however, a brief review of the more common approaches is given here.

5.5.1 Preference mapping

Preference mapping refers to a range of multivariate techniques that illustrate, using perceptual maps, the relationship between products, their sensory attributes and consumer liking. Typically, the sensory attributes of a range of products are evaluated by trained assessors and the same products are evaluated by consumers for liking. In addition to mapping individual consumers, preference mapping can also map consumer segments exhibiting similar liking. Such segments are often identified by cluster analysis of the consumer data prior to preference mapping. Different types of preference mapping exist and these can be divided into internal preference and external preference mapping. More recently partial least squares (PLS) regression has also been used as a preference mapping technique. All can be applied to the same data set but differences exist in how products and consumers are projected onto the final preference maps. Preference mapping can be applied to determine sensory drivers of a product category, identify target sensory based consumer segments and identify product opportunities.

For more detailed information on preference mapping see MacFie (2007).

5.5.2 Conjoint analysis

This technique is used to determine the relative impact of different product attributes. These can be sensory attributes but, more often than not, they are combined with other features such as price, packaging or even the context in which they could be consumed. Conjoint analysis can be applied in product design and optimisation. A predesigned set of products are presented to consumers, according to an appropriate statistical design, who are then asked to choose between or score particular 'attribute' combinations for liking, likelihood of purchase, etc. The data are then used to build models of consumer choice that can be used to predict consumer behaviour and/or identify the optimum combination of attributes for a product.

For more detailed information on conjoint analysis see Moskowitz et al. (2006).

5.5.3 Other modelling techniques

There are several other regression modelling techniques that enable the optimisation of products by modelling sensory properties or liking from chemical, physical process and/or sensory data, including response surface modelling (RSM), PLS and path PLS. Such advanced techniques require the input of a statistician and the reader should consult more advanced text for further information.

6 Completing the project

6.1 Reporting

Reporting serves two functions: to communicate findings and provide a record of the study. It is important to choose the appropriate format(s) for reporting, taking into account the audience, e.g. technical, non-technical, marketing, R&D, and circumstances. It may be necessary to use several reporting formats for one study, e.g. oral presentation and written report. Formats include the following:
- *A full written report*: This provides a full, detailed record of the study.
- *A short written report*: This format saves time and is useful when reporting routine tests, in which only deviations from the standard methodology need to be noted, and/or when the report reader is fully conversant with the test and requires only objectives, results and interpretation.
- *Scientific paper*: This is intended for an academic audience and the format is specified by the publishing journal.
- *Oral presentation*: It is important to use good oral presentation techniques and include the appropriate level of detail. Oral presentations are often accompanied by slides, which may be given to the audience as a written record and may include more detail than presented orally.
- *Poster*: The study is visually displayed and can be presented orally to people visiting the display. This reporting format is often used at conferences, exhibitions and in-house wall displays.

Reports often provide a basis for future work. It is important that they are retrieved in relevant searches and, therefore, careful consideration needs to be given to the title, key words, authors and contact details.

Reports often contain confidential information. Consideration needs to be given to the level of security and access assigned to the report. For example, it is typically not appropriate to allow confidential company information to be reported in the public domain. Careful consideration needs to be given to reporting of patentable information (see Section 4.2.6). Personal information about assessors should not be reported in a manner that allows identification of individuals, unless specific permission has been obtained.

It is also essential to include acknowledgement of copyrights and trademarks and to give credit, through referencing, to sources of information used, particularly when publishing reports in the public domain, to prevent litigation. Local copyright laws must be adhered to and permission sought to include information used verbatim from other sources, as appropriate.

By convention, a report should be written in the third person and in the past tense. The key components of reporting are listed below, although not all elements are necessary for every format.

- *Title*: It should be concise, relevant and specific. As the title is often used to search against, it should contain key words.
- *Authors*: Authors' names and contact details. It may be necessary to include institute or departmental contact details, if a long-term contact is needed.
- *Summary/abstract*: This should provide a one-page overview of the project, including objectives, key findings and recommendations. It should give enough information to provide an overview of the project without referring to the main text.
- *Objectives*: This includes clear statement of objectives or aims of the study.
- *Action standard*: This is a clear, unequivocal statement of requirements necessary for action to be taken (see Section 3.6).
- *Background/Introduction*: This should include why the project was undertaken; the broader context of the project, such as commercial context, scientific state of knowledge, whether it is part of an ongoing research program; the scope of the project; constraints and their implications; and advantages or limitations of the approach.
- *Timings*: This may include dates when testing occurred and timelines.
- *Method*: This is a clear outline of methodology including the following:
  - Experimental design, e.g. design, number and type of replications and so on.
  - Sensory method, e.g. type of test, scales used, data collection method and so on.
  - Assessors, e.g. type, number, screening, selection, level of training, instructions given and so on.
  - Samples, e.g. formulation, coding, preparation, serving and so on.
  - Test conditions, e.g. test location, environment and so on.

- Data handling, e.g. coding, transformations, treatment of missing data and so on.
- Data/statistical analysis techniques.
* *Results and Discussion*: These include the following:
  - Concise summaries of the data.
  - Tables and figures (graphs, photos, diagrams, etc.) to illustrate data. These are often given in appendices. The same data should not be presented twice. All tables and figures should be labelled consecutively and cited in the text.
  - Probability level(s), degrees of freedom, test statistic(s) and, where appropriate, the direction of the effect.
  - Logical interpretation of results with regard to the experimental design.
* *Conclusions and Recommendations*: Any conclusions and recommendations for action to be taken, as a result of the study, should be clear and relate back to the objectives and action standards.
* *References*: Typically, references are cited in the text using numbering, or authors and year; full references are listed in a reference section at the end of the report or included as footnotes. The reference format may be defined by the organisation/publisher. References provide further supporting information, further reading and often give useful leads when conducting broader literature searches. They should be relevant and appropriate to the study.
* *Bibliography*: This is for additional sources of supplementary information such as general text books and websites.
* *Appendices*: These are used for information, such as tables and questionnaires, that would detract from the flow of the main text.

6.2 Documentation and data storage

It is important that all documentation (sample information, summary data, interim and final reports, ethics applications, contact information, email communications, etc.) and raw data relating to the project are archived in an organised manner, and stored securely so as to minimise deterioration, e.g. through dampness, pests and so on. It is prudent to have multiple copies of electronic data. Records should be easily retrievable when required; they may be needed to answer additional questions or to be combined with data collected in future studies.

Raw data should be stored in its original format, e.g. electronic files, questionnaires, test ballots, lab books and so on. Personal data should be stored securely according to local data protection legislation. Some legislation gives individuals the right to request a copy of personal information held about them that must be delivered to them within a certain time period.

There are guidelines governing the length of retention of certain types of data. This may vary according to quality systems, codes of practice or legal requirements, and historically may vary between 5 and 30 years. At a minimum, data and related documentation should be kept until the report is approved.

6.3 Dos and don'ts

✔ Plan time in the project to complete a formal report. It is all too easy to move on to the next project without capturing information that could prove valuable in the long term.
✔ Ensure that the report can be easily retrieved from searchable databases, e.g. title, authors, key words, and so on.
✔ Write the abstract carefully, comprehensively and concisely. It may be the only part of the report that is read, particularly by senior management.
✔ Ensure conclusions relate back to objectives.
✔ Do report sufficient information on how the sensory test was carried out to enable the study to be replicated.
✔ Plan time to organise and archive data and documentation.
✔ Store personal records securely and in accordance with local data protection legislation.
✔ Observe local regulations governing copyright.
✔ Observe company policy regarding confidentiality.

7 Appendices

Appendix 1 Examples of Latin Square and Williams Latin Square designs for selected number of samples

Latin Square designs

5 × 5

A B C D E
B C D E A
C D E A B
D E A B C
E A B C D

6 × 6

A B C D E F
B C D E F A
C D E F A B
D E F A B C
E F A B C D
F A B C D E

Williams Latin Square designs

Size = 5

A B E C D
B C A D E
C D B E A
D E C A B
E A D B C
D C E B A
E D A C B
A E B D C
B A C E D
C B D A E

Size = 7

A B G C F D E
B C A D G E F
C D B E A F G
D E C F B G A
E F D G C A B
F G E A D B C
G A F B E C D
E D F C G B A
F E G D A C B
G F A E B D C

See MacFie *et al.* (1989) for further examples of Williams Latin Square Designs.

Appendix 2 IFST PFSG professional code of conduct for sensory professionals

(Printed with permission from IFST.)

Ethical and Professional Practices for the Sensory Analysis of Foods*

The Institute of Food Science & Technology, on the advice of its Professional Food Sensory Group, has authorised the following Statement, issued in November 2008.

These guidance notes have been drawn up by the Professional Food Sensory Group of the Institute of Foods Science and Technology, and are designed to cover the use of the techniques of sensory analysis or sensory evaluation of food, beverage and ingredients in research or quality control. They are not designed to cover the use of the techniques for large-scale surveys, for which the guidelines from the Market Research Society should be referred to (Ref. 1). These principles can also be extended to non-foods, including fragrances and products for which skin absorption can occur. Testing with children should be carried out with reference to Guidelines for Research issued by the National Children's Bureau (Ref. 2).

The principles described below should be given full consideration in the design and execution of sensory tests.

1 General Principles

1.1 The scope of permitted tests using human subjects, and levels of authorisation to sanction tests, should be defined in a written Organisation Ethical Policy.

1.2 All test procedures should be carried out in such a way as to reduce any risks to the health of the participants, whether Organisation employees, trained external assessors or consumers.

1.3 Test participants should be volunteers, either through contractual agreement or on an *ad hoc* basis, and should be able to withdraw from the testing at any time, without having to give reasons.

2 Specific Issues

2.1 The Organisation Ethical Policy should be drawn up with reference to the ACNFP (Advisory Committee on Novel Foods and Processes) Guidelines on the conduct of taste trials involving novel foods or foods produced by novel processes (Ref. 3). The principle underlying

---
* The term "foods" will be used to cover all foods, beverages and ingredients.

these Guidelines is that *"those carrying out the trial are satisfied, after taking suitable professional advice, that it poses no hazard to human health".* The Policy will depend on the nature of each individual organisation, but should typically comprise an internal mechanism to define and monitor ethical procedures, together with expert input from external sources, where appropriate.

2.2 All tests should be subject to a basic risk assessment. These will include sensory tests carried out on both standard and non-standard foods, as defined below:
- Standard foods manufactured, stored and prepared under commercial and approved conditions, and which are unlikely to need any specific requirements.
- Non-standard foods might comprise: foods containing ingredients that are not approved in the country in which the test is carried out; foods produced using novel processes; ingredients not normally consumed unless incorporated into foods; food stored under non-standard conditions; and foods containing pharmacologically active ingredients. Risk assessments should be made with reference to 2.6.

2.3 Those working with food should have received adequate training in food hygiene, and this is the responsibility of the employer/agency. NOTE: the Level 2 award in Food Safety in Catering provides suitable training in basic food hygiene.

2.4 Assessors should give informed consent to tests on non-standard foods, and should be allowed to withdraw from the panel at any time, without penalty or having to give a reason. The work should be described in such detail as is appropriate, and any information that might be relevant to possible unidentified hazards should be explained. This is particularly relevant in, for example, sensory Quality Control testing, in which there is a small but finite risk of unknown hazards. Informed consent must be given to tests on non-standard foods, and with reference to the ACNFP Guidelines.

2.5 Recording of data on assessors should be in accordance with the provisions of relevant data protection legislation of the country concerned.

2.6 Potential adverse effects on the health of assessors should be minimised, specifically:
- Panel recruitment procedures should be designed to identify known health problems and allergies, and to exclude individuals at risk.
- Products should be microbiologically safe, and if necessary the tests should be approved by a food microbiologist or should be

Appendices 145

subject to microbiological testing. This is particularly important for shelf-life and accelerated shelf life testing.
- Ingredient lists (in accordance with food labelling regulations) should be available for assessors. All foods that are under development, and which do not have standardised ingredient lists, should be subject to a risk assessment.
- Tests should be designed to minimise the amount consumed for health and nutritional reasons. In particular, tests on ingredients alone or in foods should consider the risks of consumption above normal levels.
- Chronic effects on health should be considered, for example in long-term testing on alcoholic beverages, or for assessors in multiple studies on different products over an extended period, such as sensory descriptive analysis. If appropriate, medical tests should be included as part of panel screening procedures. Records of consumption should be kept and health monitored on an ongoing basis.

2.7 Tests must be monitored for any adverse reactions, and emergency procedures must be in place in the event of such reactions. Assessors participating in long-term and multiple studies must be monitored for any developing adverse effects.

2.8 Misleading of assessors should be minimised. It is sometimes necessary to mislead assessors as to the nature of the samples or of the experiment, but this must be clearly justified and the reasons recorded in advance of the experiment.

2.9 Pain, distress or discomfort to assessors should be avoided if possible. If significant pain, distress or discomfort is involved, the assessors should be warned, and local ethical approval should be sought. In particular, invasive procedures should be minimised. If for example the use of anaesthetics on the tongue is proposed, or non-clinical x-ray, medical advice should be taken and this should be consistent with the Organisation Ethical Policy.

2.10 Testing with vulnerable groups can necessitate specific considerations. For example:
- Parental approval should be given for testing with children below the age of 16 (Ref 1).
- Testing with elderly people requires care and discretion in order to reduce the risk of inadvertent intimidation.
- Testing with people likely to have impaired immune systems and other medical conditions, including learning disabilities, requires medical advice.

146 Sensory evaluation

3 References

[1]MRS, http://www.marketresearch.org.uk/code.htm
[2]NCB,http://www.ncb.org.uk/dotpdf/open%20access%20-%20phase%201%20only/research_guidelines_200604.pdf
[3]ACNFP, http://www.acnfp.gov.uk/acnfppapers/inforelatass/guidetastehuman/guidetaste

The Institute of Food Science & Technology (IFST) is the independent professional qualifying body for food scientists and technologists. It is totally independent of government, of industry, and of any lobbying groups or special interest groups. Its professional members are elected by virtue of their academic qualifications and their relevant experience, and their signed undertaking to comply with the Institute's ethical Code of Professional Conduct. They are elected solely in their personal capacities and in no way representing organisations where they may be employed. They work in a variety of areas, including universities and other centres of higher education, research institutions, food and related industries, consultancy, food law enforcement authorities, and in government departments and agencies. The nature of the Institute and the mixture of these backgrounds on the working groups drafting IFST Statements and Guidelines, and on the Committees responsible for finalising and approving them, ensure that the contents are entirely objective.

IFST recognises that research is constantly bringing new knowledge. However, collectively the profession is the repository of existing knowledge in its field. It includes researchers expanding the boundaries of knowledge and experts seeking to apply it for the public benefit. Its purposes are:
- *to serve the public interest by furthering the application of science and technology to all aspects of the supply of safe, wholesome, nutritious and attractive food, nationally and internationally;*
- *to advance the standing of food science and technology, both as a subject and as a profession;*
- *to assist members in their career and personal development within the profession;*
- *to uphold professional standards of competence and integrity.*

The Institute takes every possible care in compiling, preparing and issuing the information contained in IFST Statements and Guidelines, but can accept no liability whatsoever in connection with them. Nothing in them should be construed as absolving anyone from complying with legal requirements. They are provided for general information and guidance and to express expert professional interpretation and opinion, on important food-related issues.

Appendix 3 Critical values table for triangle test

Minimum numbers of correct responses to reject the null hypothesis of 'no difference' at selected significance levels with a total number of assessors 'n'.

n	Significance (%)						n	Significance (%)					
	30	20	10	5	1	0.1		30	20	10	5	1	0.1
5	3	4	4	4	5	–	42	17	18	19	20	22	25
6	3	4	5	5	6	–	48	19	20	21	22	25	27
7	4	4	5	5	6	7	54	21	22	23	25	27	30
8	4	5	5	6	7	8	60	23	24	26	27	30	33
9	4	5	6	6	7	8	66	25	26	28	29	32	35
10	5	6	6	7	8	9	72	27	28	30	32	34	38
11	5	6	7	7	8	10	78	29	30	32	34	37	40
12	5	6	7	8	9	10	84	31	33	35	36	39	43
13	6	7	8	8	9	11	90	33	35	37	38	42	45
14	6	7	8	9	10	11	96	35	37	39	41	44	48
15	6	8	8	9	10	12	102	37	39	41	43	46	50
16	7	8	9	9	11	12	108	40	41	43	45	49	53
17	7	8	9	10	11	13	114	42	43	45	47	51	55
18	7	9	10	10	12	13	120	44	45	48	50	53	57
19	8	9	10	11	12	14	126	46	47	50	52	56	60
20	8	9	10	11	13	14	132	48	50	52	54	58	62
21	8	10	11	12	13	15	138	50	52	54	56	60	64
22	9	10	11	12	14	15	144	52	54	56	58	62	67
23	9	11	12	12	14	16	150	54	56	58	61	65	69
24	10	11	12	13	15	16	156	56	58	61	63	67	72
25	10	11	12	13	15	17	162	58	60	63	65	69	74
26	10	12	13	14	15	17	168	60	62	65	67	71	76
27	11	12	13	14	16	18	174	62	64	67	69	74	79
28	11	12	14	15	16	18	180	64	66	69	71	76	81
29	11	13	14	15	17	19							
30	12	13	14	15	17	19							
31	12	14	15	16	18	20							
32	12	14	15	16	18	20							
33	13	14	15	17	18	21							
34	13	15	16	17	19	21							
35	13	15	16	17	19	22							
36	14	15	17	18	20	22							

Note 1: The values in this table were calculated from the exact binomial law formula for parameter $p = 1/3$ with n repetitions (responses).

Note 2: When the number of responses is larger than 100 ($n > 100$), use the following formula to calculate the minimum number of responses (X) required to reject the 'no difference' hypothesis.

X is the nearest whole number to the value given by the following formula.

$$X = 0.4714 z \sqrt{n} + \frac{2n+3}{6}$$

where

$Z = 1.282$ for $p \leq 0.1$

$Z = 1.645$ for $p \leq 0.05$

$Z = 2.326$ for $p \leq 0.01$

$Z = 3.090$ for $p \leq 0.001$

Appendix 4 Critical values table for duo-trio test and paired comparison test for difference (one tailed)

Minimum numbers of correct responses to reject the null hypothesis of 'no difference' at selected significance levels with a total number of assessors 'n'.

n	Significance (%)			n	Significance (%)		
	5	1	0.1		5	1	0.1
5	5	–	–	29	20	22	24
6	6	–	–	30	20	22	24
7	7	7	–	31	21	23	25
8	7	8	–	32	22	24	26
9	8	9	–	33	22	24	26
10	9	10	10	34	23	25	27
11	9	10	11	35	23	25	27
12	10	11	12	36	24	26	28
13	10	12	13	37	24	26	29
14	11	12	13	38	25	27	29
15	12	13	14	39	26	28	30
16	12	14	15	40	26	28	30
17	13	14	16	41	27	29	31
18	13	15	16	42	27	29	32
19	14	15	17	43	28	30	32
20	15	16	18	44	28	31	33
21	15	17	18	45	29	31	34
22	16	17	19	46	30	32	34
23	16	18	20	47	30	32	35
24	17	19	20	48	31	33	36
25	18	19	21	49	31	34	36
26	18	20	22	50	32	34	37
27	19	20	22	52	33	35	38
28	19	21	23	56	35	38	40

Sensory evaluation

n	Significance (%)			n	Significance (%)		
	5	1	0.1		5	1	0.1
60	37	40	43	**84**	51	54	57
64	40	42	45	**88**	53	56	59
68	42	45	48	**90**	54	57	61
70	43	46	49	**92**	55	58	62
72	44	47	50	**96**	57	60	64
76	46	49	52	**100**	59	63	66
80	48	51	55				

Appendix 5 ANOVA explained

ANOVA can be applied to many different types of data. In this book, we focus on its application to sensory data.

Purpose

ANOVA is used to examine the different sources of variation within a data set, e.g. variation from different types of products or different assessors. ANOVA calculates an F-ratio to determine if these sources of variation are significantly greater than that due to background noise (error).

Consider Figure A5.1 which shows two data sets A and B. Within each data set, there are three samples with the same mean value for a sensory attribute. The variation within each sample is due to background noise and is larger in A than in B. It is easy to conclude that there is a significant difference between samples in B because the distance between mean values is obviously larger than the variation within the samples. In A, however, the variation within each sample is sufficiently large that, without ANOVA, it is impossible to determine if a significant difference exists between the samples.

Sources of variation

In sensory testing, the main sources of variation relate to samples and assessors, and potential interaction between the two. It is important to consider all sources of potential variation during the experimental design, so that they can be accounted for in the data analysis.

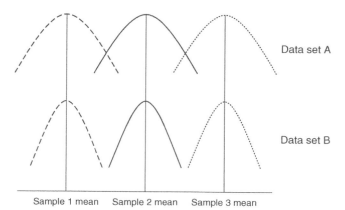

Figure A5.1 Illustration of sample mean and variation in two different data sets (A and B).

152 Sensory evaluation

Samples: Measuring variation due to samples, and identifying significant differences between them, is often the main objective of ANOVA. In order to measure the variation due to samples, the experimental design must include multiple responses for each sample.

Assessors: Variation due to assessors can arise from different use of the rating scale. For example, some individuals have a natural tendency to use the upper or lower end of the scale. In order to measure the variation due to assessors, the experimental design must allow all samples to be evaluated by all assessors. This is known as a repeated measures design.

Interaction: Variation due to the interaction between the samples and assessors can arise when different assessors place the samples in different orders of perceived intensity (crossover interaction), or when the relative magnitude of differences between samples is inconsistent across assessors (magnitude interaction). In order to measure the variation due to interaction, the experimental design must include replicated responses from each assessor for each sample.

ANOVA designs

ANOVA designs relate to the experimental design for the study. The complexity of the ANOVA is determined by the structure of the data set. The three most common designs used for sensory data are as follows:

1 *One-factor ANOVA*: This calculates variation due to one factor only; all other variation is considered as background noise (error). Commonly, the factor is the samples, where a different group of assessors provide results for each sample. The ANOVA compares the variation between sample means to variation within the samples. Alternatively, the factor could be assessors, where one sample is assessed on multiple occasions by each assessor. Here the ANOVA compares the variation between assessor means to variation within the assessors.

2 *Two-factor ANOVA (repeated measures)*: This calculates variation due to two factors; any other variation is considered as background noise (error). Typically, the factors are samples and assessors, where each assessor has evaluated each sample. The ANOVA compares variation between samples means to variation within samples that cannot be accounted for by variation across assessors. (This is the key difference between one- and two-factor ANOVA: in one-factor ANOVA, any variation due to assessors cannot be quantified and, therefore, remains in the background noise (error).) The ANOVA also compares variation between assessors to variation within assessors that cannot be accounted for by variation across samples.

3 *Two-factor ANOVA with interaction*: This calculates variation due to two factors and any interaction between these two factors; any other variation is considered as background noise (error). Typically, the factors are samples and assessors, where each assessor has provided replicated responses for each sample. The ANOVA compares variation between samples' means to variation within samples that cannot be accounted for by variation across assessors, and interaction between samples and assessors. This is the key difference between two-factor ANOVA with and without interaction: in two-factor ANOVA (without interaction), any variation due to interaction cannot be quantified and, therefore, remains in the background noise (error). The two-factor ANOVA with interaction also compares variation between assessors to variation within assessors that cannot be accounted for by variation across samples and interaction between samples and assessors.

Note: If you have chosen an ANOVA that does not calculate a specific source of variation, *do not* assume that this variation does not exist. Its effect will be contained in the background noise (error).

Calculating ANOVA

Whilst it is possible to calculate ANOVA by hand, nowadays, it is typically performed using computer software. It is important, however, to understand the origins of each term in the software output and to interpret them correctly. Table A5.1 shows the key elements of an ANOVA output for a two-factor ANOVA with interaction.

Interpretation

ANOVA output is interpreted through reference to the p-values associated with samples, assessors and interaction. If the p-value is less than the specified significance level (typically < 0.05), it can be concluded that this factor has a significant effect. For example, in Table A5.1, there is a significant effect of samples, assessors and the interaction between the two for the attribute hardness. ANOVA does not identify where significant differences within a factor exist. For example, there may be a significant sample effect but ANOVA does not identify which samples are significantly different to one another. This requires further analysis using an MCT.

Multiple comparison tests

MCTs identify the levels (samples or assessors) within a factor, between which significant differences exist. They should be applied only when the total effect for the ANOVA is significant. Note that carrying out multiple t-tests is not appropriate as it does not allow for any adjustment of the

154 Sensory evaluation

Table A5.1 ANOVA output for hardness

	SS	df	MS	F-ratio	P
Total	1094.79	62	17.658	32.89	<0.0001
Sample	1034.18	6	172.36	321.01	<0.0001
Assessor	18.99	8	2.37	4.42	<0.0001
Sample*assessor	41.62	48	0.87	1.62	0.018
Error	67.655	126	0.537		

SS: Sum of squares. The measure of variation within the data set.
df: Degrees of freedom relates to the number of levels within each element of the data set.
MS: Mean square. This is a measure of variation that takes into account the number of levels within each element of the data set. It is calculated by dividing the sum of squares by the df.
F-ratio: This is calculated, for each factor, by dividing the mean square for that factor by the mean square for the error term (background noise). This is how ANOVA compares the variation due to factors to variation from the background noise (error). The F-ratio is used to determine if there is a significant effect of each factor. If calculating ANOVA by hand, this F-ratio would be compared to a critical value from statistical tables at a specified significance level. Some software output also includes the critical F-value.
p: The p-value is a calculated probability. It is the probability of making a type I error, i.e. concluding that a significant effect exists when it does not. This relates to the significance level of the test decided at the planning stage (typically 5% for sensory test); therefore, a p-value of <0.05 would indicate a significant effect.
Note 1: Output from a one-factor ANOVA would only include total, sample (or assessor) and error terms. Two-factor ANOVA (without interaction) would include total, sample, assessor and error terms.
Note 2: Remember that an underlying assumption of ANOVA is that the data are normally distributed. The data should be checked for normality before carrying out ANOVA. If it is not normally distributed, appropriate, alternative analyses should be applied.

α risk. It may lead to the conclusion that differences exist when it is not the case.

Most MCTs work by comparing the difference between the mean values of all possible pairs of samples or assessors to a calculated value or range of values. If the difference between two sample means is greater than the calculated value, then a significant difference is concluded to exist.

Choice of multiple comparison test

There are several MCTs to choose from, each calculated in a different way (see O'Mahony 1986). The choice of test should be made prior to the analysis and is directed by the specific objective of the investigation.

Table A5.2 Results from Tukey's HSD test on sweetness for six confectionary samples

Sample	Mean				
CF53	3.5	A			
CF78	3.7	A			
CF81	4.2	A	B		
CF22	4.9		B		
CF15	5.8			C	
CF48	6.7				D

The most common tests employed are the LSD test, the Newman Keuls test and the Tukey's honestly significant different (HSD) test.

The choice of test generally depends on the risk you are willing to take in terms of missing differences that actually exist (or concluding that a difference exists when it does not) – the conservativeness of the test. Some MCTs adjust the significance level so that it is kept at 0.05 (or 0.01), for comparisons made between individual pairs of samples. These tests, e.g. Newman Keuls, Duncan and the LSD test, are more likely to find differences between pairs of samples, i.e. they are less conservative. The LSD test is the least conservative test and should be used only when there are a small number of comparisons to make, e.g. between three or four samples/assessors.

Other tests, such as the Tukey HSD, Sheffe and Bonferroni tests, work by keeping the overall level of significance for the whole set of comparisons at 0.05 (or 0.01). They are more conservative and so may miss real differences between pairs of samples.

A typical output from an MCT on a set of six samples is shown in Table A5.2.

The table lists the mean scores associated with each sample and a letter code. Samples with the same letter codes are not significantly different. Samples with different letter codes are significantly different. In the earlier example, samples CF53, CF78 and CF81 are not significantly different. Sometimes a sample is associated with more than one letter code. In the earlier example, sample CF81 is also not significantly different to CF22 but samples CF53 and CF78 are. Samples CF15 and CF48 are significantly different to each other, and the other samples.

Other MCTs exist for specific situations and include Dunnett's test, used when individual samples are compared to one sample only, e.g. a control, and Dunn's test, which is used when only selected pairs of samples are identified for comparison prior to the ANOVA.

See O'Mahony (1986) and Lea et al. (1997) for more detailed information on ANOVA and MCTs.

Appendix 6 Critical values table for chi-squared

The table lists the critical values of chi-square for degrees of freedom shown in the left-hand column for tests corresponding to those significance levels heading each column. If the calculated value for χ^2 is greater than or equal to the tabled value, reject the null hypothesis.

df	Significance (%) for one-tailed test					Significance (%) for one-tailed test			
	5	2.5	0.5	0.05		5	2.5	0.5	0.05
	Significance (%) for two-tailed test								
	10	5	1	0.1		10	5	1	0.1
1	2.71	3.84	6.64	10.83	22	30.81	33.92	40.29	48.27
2	4.60	5.99	9.21	13.82	23	32.01	35.17	41.64	49.73
3	6.25	7.82	11.34	16.27	24	33.20	36.42	42.98	51.18
4	7.78	9.49	13.28	18.46	25	34.38	37.65	44.31	52.62
5	9.24	11.07	15.09	20.52	26	35.56	38.88	45.64	54.05
6	10.64	12.59	16.81	22.46	27	36.74	40.11	46.96	55.48
7	12.02	14.07	18.48	24.32	28	37.92	41.34	48.28	56.89
8	13.36	15.51	20.09	26.12	29	39.09	42.69	49.59	58.30
9	14.68	16.92	21.67	27.88	30	40.26	43.77	50.89	59.70
10	15.99	18.31	23.21	29.59	32	42.59	46.19	53.49	62.49
11	17.28	19.68	24.72	31.26	34	44.90	48.60	56.06	65.25
12	18.55	21.03	26.22	32.91	36	47.21	51.00	58.62	67.99
13	19.81	22.36	27.69	34.53	38	49.51	53.38	61.16	70.70
14	21.06	23.68	29.14	36.12	40	51.81	55.76	63.69	73.40
15	22.31	25.00	30.58	37.70	44	56.37	60.48	68.71	78.75
16	23.54	26.30	32.00	39.29	48	60.91	65.17	73.68	84.04
17	24.77	27.59	33.41	40.75	52	65.42	69.83	78.62	89.27
18	25.99	28.87	34.80	42.31	56	69.92	74.47	83.51	94.46
19	27.20	30.14	36.19	43.82	60	74.40	79.08	88.38	99.61
20	28.41	31.41	37.57	45.32					
21	29.62	32.67	38.93	46.80					

Appendix 7 Critical values table for paired comparison and paired difference test (two tailed)

Minimum numbers of correct responses to reject the null hypothesis of 'no difference' or 'no preference' at selected significance levels with a total number of assessors 'n'.

n	Significance (%)			n	Significance (%)		
	5	1	0.1		5	1	0.1
5	–	–	–	31	22	24	25
6	6	–	–	32	23	24	26
7	7	–	–	33	23	25	27
8	8	8	–	34	24	25	27
9	8	9	–	35	24	26	28
10	9	10	–	36	25	27	29
11	10	11	11	37	25	27	29
12	10	11	12	38	26	28	30
13	11	12	13	39	27	28	31
14	12	13	14	40	27	29	31
15	12	13	14	41	28	30	32
16	13	14	15	42	28	30	32
17	13	15	16	43	29	31	33
18	14	15	17	44	29	31	34
19	15	16	17	45	30	32	34
20	15	17	18	46	31	33	35
21	16	17	19	47	31	33	36
22	17	18	19	48	32	34	36
23	17	19	20	49	32	34	37
24	18	19	21	50	33	35	37
25	18	20	21	52	34	36	39
26	19	20	22	56	36	39	41
27	20	21	23	60	39	41	44
28	20	22	23	64	41	43	46
29	21	22	24	68	43	46	48
30	21	23	25	70	44	47	50

Sensory evaluation

n	Significance (%)			n	Significance (%)		
	5	1	0.1		5	1	0.1
72	45	48	51	90	55	58	61
76	48	50	53	92	56	59	63
80	50	52	56	96	59	62	65
84	52	55	58	100	61	64	67
88	54	57	60				

Appendix 8 Critical values table for Friedman test

The table lists the critical values for Friedmans test. Different numbers of assessors are shown in the left hand column and different numbers of samples are shown at the head of each column. The table also includes two different significance levels for the test. If the calculated value for 'T' is greater than or equal to the tabled value, reject the null hypothesis.

Number of assessors (J)	Number of samples (products) (P)					
	3	4	5	3	4	5
	$p = 0.05$			$p = 0.01$		
2		6	7.6			8
3	6	7	8.53		8.2	10.13
4	6.5	7.5	8.8	8	9.3	11
5	6.4	7.8	8.96	8.4	9.96	11.52
6	6.33	7.6	9.49	9	10.2	13.28
7	6	7.62	9.49	8.85	10.37	13.28
8	6.25	7.65	9.49	9	10.35	13.28
9	6.22	7.81	9.49	8.66	11.34	13.28
10	6.2	7.81	9.49	8.6	11.34	13.28
11	6.54	7.81	9.49	8.9	11.34	13.28
12	6.16	7.81	9.49	8.66	11.34	13.28
13	6	7.81	9.49	8.76	11.34	13.28
14	6.14	7.81	9.49	9	11.34	13.28
15	6.4	7.81	9.49	8.93	11.34	13.28

With more assessors, T approximately follows a χ^2 distribution and the appropriate table can be consulted for $P - 1$ degrees of freedom.

Appendix 9 Types of scales

The scale will be selected by the sensory analyst to meet the study objectives. Different scales may need different statistical treatments (see Section 3.8). Examples of some commonly used scales are as follows (see also Figure A9.1).

Line scale (visual analogue scale)

A line scale is a continuous horizontal or vertical straight line that may be plain (unstructured) or have marks (structured), e.g. mid-point, indented end-of-scale marks. Assessors place a mark on the scale to indicate perceived intensity, which is then converted into a number representing the distance of the mark from the zero end of the scale. The scale yields interval data and parametric statistics can be used to analyse them (as long as the data are normally distributed). Line scales are typically 15 cm horizontal lines, with vertical marks or anchors at the ends or 1.5 cm in from the ends. Indenting the end anchors is said to reduce end-of-scale effects. It is generally accepted that a structured line scale with 5–10 marks behaves as a category scale.

Category scale

A scale of discrete response alternatives, e.g. words, numbers, compartments, on a scale. The ratings on a category scale are converted into numbers for further analysis. Categories, however, cannot be assumed to be equal perceptual distances apart, although the conversion to numbers makes them appear numerically equal distances apart. Hence, the scale yields ordinal data and nonparametric statistics should be used to analyse data from them (see Section 3.8.7.1).

Magnitude estimation

Intensity is assigned as a ratio, e.g. twice as strong as x. Assessors or the sensory scientist assign a value to the first sample and all subsequent samples are rated relative to it, or each sample is rated relative to the sample preceding it. Alternatively, the sensory scientist can provide a separate reference sample as the fixed modulus against which all samples are compared. The scale is infinitely long, so end-of-scale bias is minimised, but training needs to be given on scale use. Furthermore, data analysis is complex. Data may need to be transformed (e.g. into logarithmic data), corrected and/or standardised dependent upon the protocol used. *For more detailed information, see ISO and ASTM standards.*

Labelled magnitude scale (category-ratio scale)

This is a labelled category scale in which the physical size of categories corresponds to the perceptual distance between categories. Labelled magnitude scale (LMS) is said to avoid end-of-scale bias, as the top end of the scale is labelled as 'strongest imaginable', which is assumed to be equitable across individuals. Caution must be exercised on determining the application of this scale, as individual assessors may use the scale differently depending on their sensitivity. For example, an assessor who is very sensitive may assign a stronger sensation to a particular category than an assessor who is insensitive, i.e. although both assessors use 'moderate', the perceived intensity each assessor is describing could vary considerably. This scale has limited use in descriptive analysis because it covers the whole perceptual range, making it difficult to discriminate small differences. The data are analysed using parametric statistics.

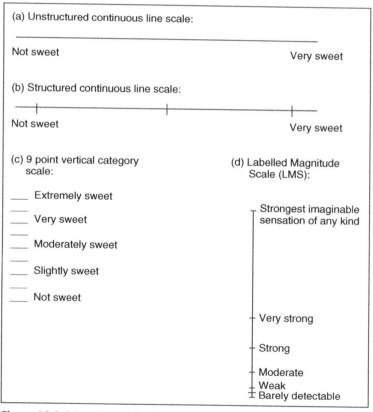

Figure A9.1 Selected examples of intensity scales used in descriptive analysis.

162 Sensory evaluation

Rank/rating

All samples are first ranked in order of perceived intensity and then rated. One method for rating is by physically placing the samples on a line scale marked on a sheet of paper. The technique can be useful for samples that are similar, attributes that are difficult to rate and assessors who are less experienced. As the technique is carried out attribute by attribute, it is more time-consuming when used in full descriptive analysis. As all the samples are ranked initially, it is not possible to combine data from additional samples at a later date. The data can be analysed based on ranks (see Section 5.2.3.3) and as appropriate for the rating scale used.

Cross-modal scaling/magnitude matching

Perceived intensity is rated by matching it to perceived intensity in another modality, e.g. odour intensity is matched to sound pitch. It avoids the use of numbers and thereby avoids the tendency of assessors to use numbers differently. There is, however, debate in the literature on whether this method produces more variable data than direct rating.

Appendix 10 Case study: modified quantitative descriptive analysis of chocolate texture

Recruitment and training

A major chocolate manufacturer wanted to set up a trained descriptive panel to measure the texture of commercial milk chocolate using a modified quantitative descriptive analysis (QDA) approach. They decided to recruit external assessors, as their permanent staff were too busy to attend regularly.

Weeks 1–4: They put an advert in the local papers describing the position and received 60 enquiries. These people were sent an application form. After screening the application forms to remove applicants who had food allergies or other medical conditions that would prevent them taking part, who would not eat confectionary, and who were unavailable to attend 2 × 2 hour session per week, 40 applicants remained.

Weeks 5–6: These 40 applicants were invited to attend one of three sessions to screen them for good sensory abilities. Only 35 applicants were able to attend. These applicants were put through a battery of screening tests, including the following:

- Recognising and detecting basic tastes, including naming tastes in solution and ranking solutions of different intensities.
- Recognising aromas related to confectionary, such as honey, caramel, burnt sugar, cocoa, milk, etc.
- Texture acuity tests, involving discriminating between different grades of sugar ranging from icing sugar to granulated sugar and naming sugar letters placed on the tongue without seeing them.
- Ability to describe sensations, by describing the sensory properties of several different chocolate-based confectionary products.
- Ability to discriminate chocolate products by carrying out several triangle tests on block chocolate with varying degrees of difference.
- Ishihara vision test to check for normal vision.
- A personal interview to assess personality, ability to work as part of a team and long-term commitment to the role.

Assessors were screened for a range of sensory abilities, as it was felt they would be used for future projects which did not just investigate texture. The results of the screening tests were assessed and 13 applicants passed the screening. Of these, 10 were offered and accepted a position as an assessor on the descriptive panel.

Weeks 7–20: Assessors then undertook a 3 month training programme that included general training and specific training on chocolate texture. All assessors attended the same sessions. Group and individual feedback was given on performance throughout the training programme. Once a month, the panel and the panel leader had an informal lunch as a group to help build the team and increase motivation.

Weeks 7–8: Assessors were given general training. This included the following:
- Sampling and group discussions on different types of chocolate-based confectionary.
- Sampling and group discussions on foods with different textural characteristics.
- Learning about scaling by rating amount of shading on different shapes.

Weeks 9–12: Assessors were then trained on chocolate texture. First, texture attributes were generated. Assessors were exposed to a set of chocolate samples that represented the extremes of textures' qualities and intensities on the market. Assessors generated terms that described the textural qualities they experienced (first column in Table A10.1).

The generated list was then revised to remove any subjective, duplicate, or ambiguous sensory terms. Upon discussion, 'graininess' and 'bittiness' were found to be duplicate terms with the same perceptual meaning; 'graininess' was retained. Softness was found to be the opposite of hardness and was removed. The final attribute list consisted of six attributes (second column in Table A10.1).

A sensory lexicon (Table A10.2) was developed and agreed by all assessors. This included the attribute name, written definition, the method of assessment and physical references that illustrated the attribute. Physical references were chocolate samples specifically manufactured to demonstrate the particular attributes.

The product assessment protocol was then determined. The overall manner of eating the chocolate, the point during the assessment when

Table A10.1 Generation of attribute list

Initial attribute list	Final attribute list
Hardness	Hardness
Crumbliness	Crumbliness
Rate of melting	Rate of melting
Thickness of melt	Thickness of melt
Softness	Graininess
Graininess	Mouth-coating
Mouth-coating	
Bittiness	

each attribute was to be rated and the order in which the attributes were to be assessed was agreed. They are as follows:
- First bite across centre of sample with incisors (assess hardness).
- Chew the portion bitten off with molars until molten.
- Swallow the sample.
- Place remaining half in mouth and hold in the mouth until it has disappeared (rate of melt).

After trying several different palate cleansers, warm water and a 1 minute break between samples were chosen. Red lighting was used to help disguise any appearance cues for texture.

Weeks 13–18: The panel were then trained to rate intensity on each attribute. First, they were trained to use the chosen scale, which was a continuous line scale with indented anchors, by rating the amount of shading on different shapes. For each attribute, assessors then ranked chocolate samples, selected to have a range of intensity along the attributes.

They worked as a group to rate the same samples on on-screen continuous line scales (marks were converted to rating out of 10). The panel leader ensured that attention was drawn to using the high end of the scale for intense samples. References, from within the sample set, were selected for each attribute to illustrate the low and high ends of the scale.

Table A10.2 Sensory lexicon

Final attribute list	Definition	Method of assessment
Hardness (not hard–very hard)	Force required to bite into chocolate	Bite through middle of sample using front teeth
Crumbliness (not crumbly–very crumbly)	Manner in which sample breaks apart when chewed	Assess on first chew following first bite on back teeth
Rate of melting (slow–fast)	Time taken to become molten chocolate	Place lump on tongue. Do not chew and wait for chocolate to melt.
Thickness of melt (not thick–very thick)	Consistency of melted chocolate	Assess once bolus is melted prior to swallowing
Graininess (not grainy–very grainy)	Presence of grains in molten chocolate	Assess molten chocolate using tip of tongue against back of front teeth
Mouth-coating (not–very)	Extent to which residue coats the mouths after swallowing	After swallowing pass tongue across surfaces of oral cavity

Assessors then worked individually to rate samples and shared their ratings with the group. This helped to calibrate the panel to produce similar ratings. Finally, assessors practiced rating samples individually in the booths with a sample set that included duplicates. Each assessor evaluated the samples in the same order, as the objective at this stage was to judge assessor performance. The group and individuals were provided with feedback, with emphasis on illustrating whether their ratings were similar for duplicate samples, which helped to make the panel consistent and repeatable.

The results of a one-way ANOVA for the panel, and each assessor, for each attribute were also discussed. For example, for crumbliness the panel were able to discriminate between the training samples ($p = 0.04$). Most assessors were also able to discriminate (p ranged from 0.01 to 0.07) except assessor 7 ($p = 0.49$). Assessor 7 realised he was assessing crumbliness too late during chewing and, on reassessing the samples, indicated similar ratings to the rest of the panel.

Unfortunately, at this point, one of the assessors resigned from the panel as he/she had found a full-time job. The company decided to continue with nine assessors, so that projects would not be delayed, and to employ one of the applicants who had passed the screening, train the applicant individually and incorporate the applicant into the panel at a later date when he/she was performing as well as the other assessors.

Weeks 19–20: A performance check was run to determine if panel performance was consistent and reliable enough on the six attributes to start working on projects. The nine assessors (1 to 9) assessed four chocolate samples (choc 1 to 4) that were representative of the full sample set, in triplicate, for the six textural attributes. The full assessment protocol, including the final ballot and sample serving protocol, was used.

Data Analysis: Data were analysed to determine the following:
- Whether the panel were sufficiently trained, i.e. whether they were consistent in their use of the scales, whether they all appeared to rate each attribute in the same way.
- Whether the panel could discriminate between the chocolate samples for all texture attributes.

Observations of histograms and tests for normality confirmed that the data were normally distributed for each attribute and sample. Figure A10.1 shows the evaluation for hardness as an example.

Consistency within assessor replicates was determined by comparing individual standard deviations (sds) with global panel sds for each sample and attribute. The majority of assessors performed consistently,

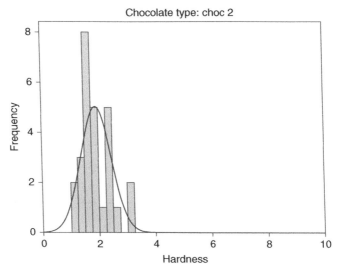

Figure A10.1 Histogram for hardness ratings for choc 2 (nine assessors, three replicates). Kolmogorov–Smirnov test confirmed that distribution was not significantly different to normality ($p = 0.141$).

i.e. within panel sd. However, this was not the case for assessor 5, whose sds were often twice the size of the global panel value. The assessor was made aware of this issue and, if no improvements are observed, he/she may need to be removed from the panel.

The data set was analysed using a two-factor ANOVA with replication and interaction. When significant differences were found *post hoc*, Tukey's HSD MCTs ($p = 0.05$) were applied.

Table A10.3 shows a typical ANOVA output for the data, using hardness as an illustration. Table A10.4 provides a summary of the level of significance associated with the factors and interaction terms in the ANOVA for each of the attributes. Table A10.5 provides the mean scores for each chocolate sample and indicates the results of the sample *post hoc* Tukey's HSD tests.

The ANOVA indicated that there was significant assessor–sample interaction for two attributes (rate of melting and graininess) suggesting that there was disagreement between the assessors. Viewing the interaction plots confirmed this. For example, Figure A10.2a displays the interaction plot for hardness showing general consistency across the panel. Ratings for choc 1 were variable but this indicated that the sample was

168 Sensory evaluation

Table A10.3 ANOVA output for hardness

	SS	df	MS	F	Significance level
Assessor	70.165	8	8.771	2.991	0.006
Sample	545.177	3	181.726	61.971	<0.001
Interaction	54.074	24	2.253	0.768	0.762
Error	211.135	72	2.932		

Table A10.4 Significance levels associated with ANOVA by attribute

	Significance level (p)		
	Assessor	Sample	Assessor*sample interaction
Hardness	0.006	<0.001	0.762
Crumbliness	0.059	<0.001	0.259
Rate of melting	0.001	0.78	<0.001
Thickness of melt	0.06	<0.001	0.055
Graininess	0.001	<0.001	<0.001
Mouth-coating	0.842	<0.001	0.086

Table A10.5 Mean attribute scores

	Choc 1	Choc 2	Choc 3	Choc 4
Hardness	5.4^a	2.1^b	6.7^c	8.1^d
Crumbliness	4.2^a	3.1^b	5.7^c	6.2^c
Rate of melting	6.2^a	5.1^a	5.2^a	5.4^a
Thickness of melt	3.1^a	5.0^b	8.0^c	5.2^b
Graininess	1.6^a	3.5^b	6.0^c	5.2^d
Mouth-coating	7.2^c	5.7^b	3.7^a	5.1^b

^{abcd}Samples with the same letter code in any row are not significantly different according to Tukey's HSD test.

not homogenous. (Subsequent analyses showed that this was the case for most attributes.)

In contrast, Figure A10.2b displays the plot for rate of melting. The crossing and differences in magnitude of the lines indicated that the assessors agreed neither on the rank order of the samples, nor on the level of rate of melting across the samples. The results showed that further discussion and training were necessary before the panel was ready to use this attribute.

Analysis of the interaction plot for graininess indicated that the interaction was a result of the data from just two assessors and, consequently, those assessors required some additional training.

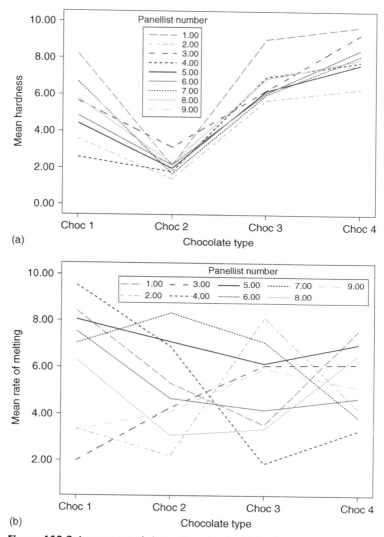

Figure A10.2 Assessor–sample interaction plots for (a) hardness and (b) rate of melting.

Significant assessor effects were found for hardness, rate of melting and graininess. For rate of melting and graininess, the significant assessor effects occurred where significant interactions were found. In these cases, the attribute qualities and/or assessment needed further discussion

Table A10.6 Results from Tukey's HSD *post hoc* test showing groupings of assessors for use of hardness scale

Assessor	Mean	Group
2	4.2	A
4	4.7	A
5	5.0	A
6	5.3	AB
7	5.6	AB
8	5.7	AB
9	5.8	AB
3	6.0	AB
1	7.2	B

as a priority. For hardness, a significant assessor effect occurred, although there was no significant interaction. Upon further examination of the data, issues concerning use of scale were apparent. Table A10.6 shows the subgroups of assessors according to Tukey's HSD test for this attribute. This indicated that assessor 1 appeared to be scoring samples at the higher end of the scale, whereas assessors 2, 4 and 5 tended to use the lower end. This was not considered to be a large variation, as differences were only within 12% of the scale. However, these assessors were given feedback to enable them to adjust their ratings in line with the panel.

Where the panel assessments were unreliable, i.e. rate of melting and graininess, it was not appropriate to comment on discrimination across the samples. It was, however, possible to conclude that the panel could distinguish between the samples for hardness, crumbliness, thickness of melt and mouth-coating.

Conclusion: The panel were not sufficiently trained to assess samples. They were, however, fairly consistent in their assessment of hardness, crumbliness, thickness of melt and mouth-coating and only two assessors needed additional training for graininess. Considerable training, however, was still required for the rate of melting attribute.

For all attributes, except mouth-coating and graininess, the panel were able to discriminate between the samples. Referring back to Table A10.5, choc 1 was particularly mouth-coating, choc 2 was not at all hard, choc 3 had a thick melt, and choc 4 was especially hard and somewhat crumbly.

Sample evaluation

After five additional training sessions, the panel were deemed ready for routine sample evaluation. Their first project was to carry out a

Table A10.7 Significance levels associated with ANOVA by attribute

	Significance level (p)		
	Assessor	Sample	Assessor*sample interaction
Hardness	<0.0001	<0.0001	0.018
Crumbliness	0.009	<0.001	0.223
Rate of melting	<0.0001	<0.0001	0.112
Thickness of melt	0.003	<0.0001	0.512
Graininess	<0.0001	<0.0001	<0.121
Mouth-coating	0.480	<0.0001	0.154

marketplace overview by assessing seven chocolate samples (A–G) (three replicates) for the six textural attributes. The results from the associated two-factor ANOVA with interactions and the PCA (with rotation) are given below. (It is important to note that using PCA is normally recommended when more attributes are evaluated and this example is simply used to illustrate key elements of interpretation.)

Table A10.7 shows that significant interaction occurred for hardness. However, on inspection of the assessor–sample interaction plot (Figure A10.3), it was evident that this interaction, although statistically significant, was of no real consequence. It was simply due to slight crossover effects and the panel were performing adequately. Most attributes indicated significant assessor terms, as is often found in the QDA approach, and, on inspection of the data, it was evident that these were caused due to slight differences in the use of the scale by some assessors and was of no consequence to the overall interpretation of the data.

Table A10.8 presents the mean attribute scores for each chocolate sample and indicates where significant differences existed according to the *post hoc* Tukey HSD test. All attributes were able to discriminate between the chocolates, with each attribute placing chocolates into at least four significantly different groups.

The PCA (with rotation) indicated that two components were able to explain 94.82% of the variation in the data. PC1 accounts for 50.4% of the variation. The correlation circles (Figure A10.4a) show that this component is highly positively correlated with graininess and thickness and negatively correlated with mouth-coating. PC2 is positively correlated with hardness and crumbliness and negatively correlated with melting rate.

172 Sensory evaluation

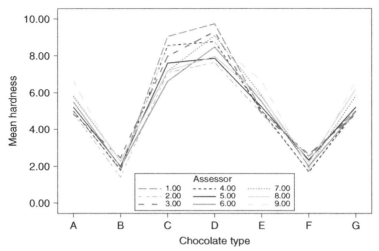

Figure A10.3 Assessor–sample interaction plots for hardness.

Table A10.8 Mean attribute scores

	Chocolate samples						
	A	B	C	D	E	F	G
Hardness	5.4^c	1.9^d	7.8^b	8.6^a	5.4^c	2.1^b	5.4^c
Crumbliness	2.8^{de}	2.3^e	6.9^b	8.3^a	5.4^c	3.4^d	2.7^e
Rate of melting	7.1^b	7.4^b	3.7^d	1.8^e	5.2^c	8.5^a	4.1^d
Melt thickness	3.1^d	5.02^c	8.1^a	5.2^{bc}	5.4^{bc}	6.1^b	3.1^d
Graininess	1.6^d	3.52^c	7.3^a	5.2^b	5.34^b	4.6^b	1.5^d
Mouth-coating	7.2^a	5.66^{bc}	3.7^d	5.1^{cd}	5.1^{cd}	6.8^{ab}	7.2^a

^{abcde}Samples with the same letter code in any row are not significantly different.

The bi-plot (Figure A10.4b) gives an overview of the texture of the chocolate sample. It characterises chocolates A and G as particularly mouth-coating in comparison to the others. Choc C is thick and grainy; choc D is particularly hard, whereas B and F are fast-melting. Choc E, however, scores mid-range for most attributes.

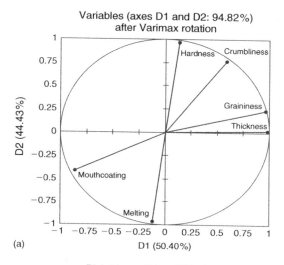

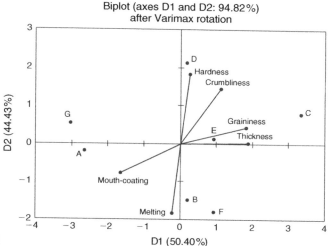

Figure A10.4 (a) Correlation circle and (b) bi-plot for principal component analysis with rotation.

Appendix 11 *R* index explained

Introduction

R index is a data analysis technique that allows the calculation of a discrimination index, i.e. how different two samples are from each other. This is sometimes preferable over the standard analyses of discrimination tests, which only determine if samples are significantly different or not.

It is a rapid technique, based on signal detection theory that can be applied to data from several different sensory methods.

Why use it?

Discrimination tests with response bias

Commonly used discrimination tests, e.g. paired comparison and triangle test, do not suffer from response bias. For paired comparison tests, the assessor is asked to indicate which sample is the strongest for a specified attribute and for triangle tests, the assessor determines which two samples are the most similar and, therefore, which is the odd one out. The mental strategy for making these judgements is quite simple and assessors with the same sensory ability would give the same response. Other methods, however, do suffer from response bias, e.g. 'A' 'not A' and same–different test. In an 'A' 'not A' test, the assessors are asked to identify a series of samples as either 'A' or 'not A'. The decision about identity will be affected by a person's willingness to take risk, i.e. how far away from 'A' do samples have to be before they are considered as 'not A'. This judgement is affected by human nature and the desire to 'not get it wrong'; one assessor may require a big difference before being willing to say 'not A' whereas another judge may say 'not A' at the slightest hint of a difference. Similarly, for the same–different test, the required 'sensory' distance between samples before they are declared as 'different' will vary among assessors. These variations do not relate to their sensory ability but, instead, are a facet of human nature and personality.

R index can be used to analyse data from 'A' 'not A' and same–different tests to provide a measure of discrimination that is free from response bias. The calculation is dependent on the use of a sureness rating with the test method, i.e. a scale that measures how sure/unsure the assessor is of his/her response. The most common sureness scales contain four or six categories.

Sure	Sure	Sure
Slightly sure	Sure?	Sure?
Very slightly sure	Sure??	Unsure?
Very slightly unsure	Unsure??	Unsure
Slightly unsure	Unsure?	
Unsure	Unsure	

A rapid technique
R index is determined by calculating the percentage of correct responses expected from a number of theoretical paired comparisons. In an 'A' 'not A' test, for example, if 10× 'A' samples and 10× 'not A' samples were presented to an assessor, this would be equivalent to 100 paired comparisons (each 'A' compared to each 'not A'). The R index would calculate the number of times 'A' would be selected in a paired comparison test. The time taken to present 20 samples would be considerably less than the time taken to present 100 paired comparisons and is far less fatiguing for the assessor. This methodology can be extended to include several different 'not A' samples, thus making it even more rapid. In this instance, a separate R index would be calculated for 'A' compared to each 'not A'. This application is particularly useful in quality control, where several different production batches (different 'not A' samples) can be compared to a standard ('A') to determine a suitable tolerance in specification.

Multiple sample assessments
As stated earlier for 'A' 'not A', R index is capable of analysing data from tests in which more than two different sample types have been presented. It can be easily applied to data from ranking tests (attribute specific and preference). Ranking tests analysed using R index can include multiple presentations of the same sample, e.g. ranking six samples for sweetness (2× sample A, 2× sample B, 2× sample C). The R index calculation uses all of the information and provides a discrimination index (degree of difference) that is more useful than identifying significant differences alone.

Calculating R index
Whichever method is used to collect the data, it is summarised in a matrix table that divides the different response categories for each sample presented.

The data within each cell of the table are used to calculate R index.

	A sure	A sure?	A sure??	A unsure??	A unsure?	A unsure
Sample A	a	b	c	d	e	f
Sample B	g	h	i	j	k	l

$$R \text{ index} = \frac{a(h+i+j+k+l)+b(i+j+k+l)+c(j+k+l)+d(k+l)+e(l)+0.5(ag+bh+ci+dj+ek+fl)}{(a+b+c+d+e+f)(g+h+i+j+k+l)} * 100$$

'A' 'not A' test

Using the 'A' 'not A' example detailed earlier, 10× presentations of each 'A' and each 'not A' sample resulted in the data shown in Table A11.1.

These results show that more 'A' samples are identified as 'A' and more 'not A' samples identified as 'not A'. The more similar the samples the more likely there will be responses in all categories for both samples.

$$\text{R index} = \frac{6(0+1+0+2+7)+2(1+0+2+7)+1(0+2+7)+1(2+7)+0(7)+0.5((6*0)+(2*0)+(1*1)+(1*0)+(0*2)+(0*7))}{(6+2+1+1+0+0)(0+0+1+0+2+7)} * 100$$

$$\text{R index} = \frac{(6(10)+2(10)+1(9)+1(9)+0+0.5(1)}{100} * 100$$

$$\text{R index} = \frac{98.5}{100} * 100$$

R index = 98.5%

Same–different test

The response matrix and calculation for the same–different test would look exactly the same. For example, 10× pairs of same samples and 10× pairs of different samples are presented in Table A11.2.

R index = 98.5%

Table A11.1 Results for 'A' 'not A' test with 10× of each sample presented

	'A' sure	'A' sure?	'A' sure??	'A' unsure??	'A' unsure?	'A' unsure
Sample 'A'	6	2	1	1	0	0
Sample 'not A'	0	0	1	0	2	7

Table A11.2 Results for same–different test with 10× of each sample pair (same or different) presented

	Same sure	Same sure?	Same Sure??	Same Unsure??	Same Unsure?	Same Unsure
Same	6	2	1	1	0	0
Different	0	0	1	0	2	7

Ranking test

For ranking tests the data are summarised in Tables A11.3 and A11.4.

$$R\ index = \frac{4(2+3+4) + 3(3+4) + 2(4) + 0.5((4*1) + (3*2) + (2*3) + (1*4))}{(4+3+2+1)(1+2+3+4)} * 100$$

$$R\ index = \frac{36 + 21 + 8 + 10}{100} * 100$$

$$R\ index = 75\%$$

How to interpret the result

Theoretically, the results of an R index indicate how many times one sample would have been selected over the other, had the two samples been presented in a paired comparison test. More typically the R index is used as a measure of discrimination, with different R indices being compared. Identical samples would yield an R index of 50%; the higher the R index value (up to a maximum of 100%) the further apart the samples. An index of 100% would result from samples that are not confusable and, as such, are not appropriate for discrimination tests.

Tables exist to determine if the R index is statistically significant (Bi and O'Mahony 2007) and, furthermore, different R index values can be analysed using ANOVA to see if any significant differences exist between them.

Table A11.3 Ranking data for four products (2× sample A and 2× sample B)

First	Second	Third	Fourth
A	A	B	B
A	B	B	A
A	B	A	B
A	A	B	B
B	A	A	B

Table A11.4 Summary of results for ranking data

	First	Second	Third	Fourth
Sample A	4	3	2	1
Sample B	1	2	3	4

The table shows the number of times each sample is ranked in a specified position.

8 Glossary

Absolute scale   A scale in which the intensity range represented by the scale is equivalent in strength across different studies.
Acceptability/acceptance test   A test to measure the degree to which a product is liked and/or favourable to consumers.
Action standard   The criteria that must be met in order to take a course of action.
Adaptation   A decrease over time in the responsiveness of the sensory system to a constant stimulus.
Affective test   A test to measure subjective consumer response.
2 Alternative force-choice test (2AFC)   An attribute-specific discrimination test in which assessors determine which of two samples has the greatest perceived intensity of a specified attribute. (Also known as a paired comparison test).
3 Alternative force-choice test (3AFC)   An attribute-specific discrimination test in which assessors determine which of three samples has the greatest perceived intensity of a specified attribute. Two samples are the same.
Analysis of variance (ANOVA)   A parametric statistical technique used to investigate the sources of variation in a data set. Typically used in sensory testing to investigate variation due to samples, assessors and other experimental variables.
Anchors   A point on a scale and/or a physical reference against which comparative judgements are made.
'A' 'not-A' test   A discrimination test for overall difference in which assessors determine if a sample is either 'A' or 'not A'. Assessors are familiarised with both products before they participate in the test.
Aroma   The sensation produced when volatile compounds stimulate olfactory receptors in the nasal cavity.
Assessor   The individual, sensory panellist, consumer, respondent, subject, and so on, giving a response.

Attribute A qualitative sensory characteristic of a product/stimulus.
Attribute diagnostic test A technique used in a consumer test to help understand the sensory basis of acceptance or liking.
Balanced A sample presentation order in which each sample occurs in each serving order position an equal number of times. See also Williams Latin Square.
Bipolar scale A scale that runs from one sensory characteristic to another rather than low to high intensity of one characteristic.
Blocks Subsets of the experimental design. These can be either samples, assessors or other design factors.
Calibration The practice of aligning sensory assessors to produce similar results carried out during the training phase.
Carrier A material with which a stimulus is normally consumed or used, but is not itself assessed, e.g. bread is a carrier for butter; skin is a carrier for fragrance.
Category-ratio scale A ratio scale with categories of response identified at measured points. (See also labelled magnitude scale.)
Category scale A scale of discrete response alternatives. The perceptual interval between each response is not necessarily equal and must be analysed using nonparametric statistics.
Category-specific scale A scale with a perceived intensity range that covers a category/product class.
Central location testing (CLT) Testing in which assessors come to one location to take part.
Chemesthesis The perception of trigeminal stimulation such as irritation, tingling and cooling initiated via chemical stimulation of sensory receptors.
Cluster analysis A group of data analysis techniques which group objects (e.g. products or assessors) into homogenous subsets based on a number of measured attributes.
Conjoint analysis A data analysis technique used to determine the relative impact of product attributes on perception, e.g. liking or purchase intent.
Consensus profiling A descriptive analysis technique in which assessors work as a group to identify qualitative sensory characteristics of a product and produce a single quantitative measure of intensity for each characteristic.
Cross-modal scaling A technique in which the perceived intensities of stimuli in one modality are matched to the perceived intensity of stimuli in another modality.

Descriptive analysis A technique for qualitatively and/or quantitatively measuring the sensory characteristics of products.

Difference from control profiling A descriptive analysis technique typically used in quality programmes in which a trained sensory panel directly measures the difference in perceived intensity of attributes of a test product to those of a reference/control product, using a degree-of-difference scale.

Difference from control test A discrimination test for overall difference in which assessors determine if a difference exists between one or more samples and a control sample, and rate the degree of difference between the sample(s) and the control.

Discrimination tests A range of techniques used to determine if a difference (or similarity) exists between two or more samples.

Discussion guide The guide used to structure and moderate the discussion in a focus group.

Dummy sample The first sample in a consumer test, the results from which are discounted.

Dumping The response bias of incorrectly assigning perceptions of attributes that are absent on the ballot to attributes present on the ballot.

Duo-trio test A discrimination test for overall difference in which assessors are asked to judge which of two samples is the same or different from a reference sample.

Enhancement The presence of one stimulus increases the perceived intensity of another. Enhancement is also defined by some as being an 'improvement' through an increase in intensity and/or liking.

Ethics committee (**Ethical Review Committee**) A recognised independent committee that determines if trials involving human subjects are ethical.

Ethnography The study of human social and cultural behaviour through direct observation.

Flash profiling A rapid, quantitative descriptive analysis technique in which assessors rank samples for individually generated attributes.

Flavour The classical definition of flavour is the total of sensations resulting from stimulation of the chemical senses in the oral and nasal cavities, namely taste, olfactory and trigeminal receptors. Flavour is defined by some as also including sensations resulting from stimulation of gustatory, olfactory, tactile, visual and auditory receptors.

Flavor Profiling® A descriptive analysis technique in which a small, trained sensory panel assess aroma, flavour and mouth-feel using a specific methodology.

Focus group   A technique in which 10–12 consumers generate qualitative information through semistructured discussion, facilitated by a moderator.

Free choice profiling   A quantitative descriptive analysis technique in which untrained assessors rate a sample set for individually generated attributes.

Friedman test   A nonparametric statistical technique used to investigate the sources of variation in a data set.

Generalised procrustes analysis (GPA)   A data analysis technique which pretreats the data to adjust for assessor variation prior to principle component analysis (PCA). Typically used to analyse data from flash and free choice profiling.

Gustation   The sense of taste.

Hedonic rating   A technique to measure the degree of liking for a product by untrained assessors.

Home use testing (HUT)   Testing in which consumers assess products in their own home.

Informed consent   The assessors giving their voluntary, duress-free consent to participate in a test, after being fully informed of the test nature, purpose, protocol, procedures and associated risks of participation.

Intensity (perceived)   Perceived strength of a stimulus.

Just-about-right (JAR) scales   A category scale used to measure the subjective response to the perceived intensity of an attribute.

Kinaesthesis   The perception of muscular body movements.

Labelled magnitude scale (LMS)   A category-ratio scale that measures perceived intensity.

Latin Square   A tool used to create balance in an experimental design.

Line scale   A continuous horizontal or vertical straight line that may be plain (unstructured) or have marks (structured).

Magnitude estimation   A ratio scaling technique in which the perceived intensity of an attribute is determined by comparison to the rating given to a reference (modulus) or the preceding sample.

Magnitude matching   See cross-modal scaling.

Modulus   A sample/stimulus with a preassigned value against which the perceived intensity of other samples can be compared.

Monadic   An experimental design in which only one sample is presented individually to each assessor during a test.

Mouth-feel   Tactile sensation perceived in the oral cavity, e.g. astringency, oily.

Multimodal perception   The integration of signals from different sensory modalities.

182 Sensory evaluation

Nonparametric tests   Statistical tests that do not make assumptions about the underlying distribution of the population or nature of the scales used to collect the data.
Odour   See aroma.
Olfaction   The sense of smell.
Paired comparison test   An attribute-specific discrimination test in which assessors determine which of two samples has the greatest perceived intensity of a specified attribute.
Paired preference test   A test in which untrained assessors identify which of two samples they prefer.
Panel   A group of assessors selected to take part in a test.
Palate cleanser   A bland food or beverage used to clear the mouth and allow sensory receptors to recover between product assessments. It can also be a time period left between samples.
Panel – consumer   A group of consumers taking part in a consumer test.
Panel leader   A trained sensory professional who is able to train a panel of assessors to generate consistent and reliable data.
Panel – sensory   A group of assessors trained to make objective sensory judgements.
Parametric tests   Statistical tests that assume that the data from the underlying populations is normally distributed.
Preference mapping   A range of multivariate techniques that illustrate, using perceptual maps, the relationship between products, their sensory attributes and consumer liking.
Principal component analysis   A data reduction technique that simplifies the visualisation of products and their attributes by representing the relationship between the original attributes on a smaller number of new variables (principle components).
Quantitative Descriptive Analysis® (QDA®)   A descriptive analysis technique in which a trained sensory panel assess a full range of sensory characteristics by generating an agreed list of attributes and individually rating perceived intensity on line scales.
Quantitative flavour profiling (QFP)   A descriptive analysis technique in which flavour characteristics are assessed by a trained sensory panel using selected terms from a predefined lexicon.
Randomisation   A tool to remove bias from experiments.
Ranking   A technique in which three or more products are placed in order of perceived intensity of an attribute or preference.
Rank-rating   A technique in which all samples in a set are first ranked in order of perceived intensity and then rated for perceived intensity.

Rating Assigning a measure to a perception.

Ratio scale A scale that has a true zero and on which ratings are proportional to one another.

Reference A sample or stimulus against which comparisons are made, or a sample representing the nature and/or intensity of sensory attribute(s).

Relative scale A scale in which the intensity range represented by the scale is related only to the attributes and products being assessed.

Response surface methodology Experimental approach that allows the simultaneous impact of two or more variables to be studied.

R index A data analysis technique used to calculate the degree of discrimination between two samples.

Same–different test A discrimination test in which assessors determine if pairs of samples are the 'same' or 'different'.

Screening The process of selecting assessors to take part in a test.

Sensitivity The ability to detect, identify or distinguish stimuli.

Sensory fatigue A decline in capability of the sensory system due to excessive stimulation or testing.

Sensory space The perceptual range covered by a sample set.

Sequential monadic An experimental design in which samples are individually presented one after another.

Somesthesis The perception of tactile sensations including temperature, pressure and pain.

Spectrum™ method A descriptive analysis technique in which a highly trained sensory panel assess a full range of sensory characteristics using a predefined, standardised lexicon.

Suppression The presence of one stimulus decreases the perceived intensity of another.

Synergy The perceived intensity of multiple stimuli is greater than the sum of the individual perceived intensities.

Temporal dominance of sensation (TDS) A technique in which trained assessors rate the perceived intensity of dominant attributes to track dominance of sensation for multiple attributes over time.

Texture The rheological, structural and geometrical properties of products perceived using tactile, visual and auditory sense organs.

Texture Profiling® A descriptive analysis technique in which a trained sensory panel assess texture and mouth-feel properties of foods using a predefined lexicon and standardised protocol.

Time intensity A technique used to measure dynamic changes in sensation over time.

Triangle test A discrimination test for overall difference in which assessors judge which of three samples is different. Two samples are the same.

Universal scale A scale that covers the full range of sensations that occur across all product classes. It is an absolute scale by definition.

Visual analogue scale Alternative name for a line scale.

Williams Latin Square An experimental design in which all samples are presented in each presentation position and, before and after every other sample, an equal number of times.

9 References

Advisory Committee on Novel Foods and Processes (ACNFP) (2000) *Guidelines on the conduct of taste trials involving novel foods or foods produced by novel processes.* http://www.acnfp.gov.uk/acnfppapers/inforelatass/guidetastehuman/guidetaste.

Anonymous (1975) *Minutes of Division Business Meeting.* Institute of Food Technologists – Sensory Evaluation Division, IFT, Chicago, IL.

Ashcroft, S. & Pereira, C. (2003) *Practical Statistics for the Biological Sciences.* Palgrave Macmillan, UK.

ASTM, Committee E-18 (1986) *Physical Requirement Guidelines for Sensory Evaluation Laboratories*, J. Eggertj & K. Zook (eds), American Society for Testing and Materials, Philadelphia, PA, ASTM Special Technical Publication 913.

ASTM Committee E-18 (1992) *Manual on Descriptive Analysis Testing for Sensory Evaluation*, R.C. Hootman (ed), American Society for Testing and Materials, Philadelphia, PA, MNL 13.

ASTM E1909-97 (2003) *Standard Guide for Time-Intensity Evaluation of Sensory Attributes.* American Society for Testing and Materials, Philadelphia, PA.

ASTM E1879-00 (2004) *Standard Guide for Sensory Evaluation of Beverages Containing Alcohol.* American Society for Testing and Materials, Philadelphia, PA.

ASTM E2299-03 (2003) *Standard Guide for Sensory Evaluation of Products by Children.* American Society for Testing and Materials, Philadelphia, PA.

Bi, J. & O'Mahony, M. (2007) Updated and extended table for testing the significance of the R-index. *Journal of Sensory Studies*, 22, 713–720.

Brace, I. (2004) *Questionnaire Design: How to Plan, Structure and Write Survey Material for Effective Market Research.* Kogan Page, Philadelphia, PA.

Brandt, M.A., Skinner, E.Z. & Coleman, J.A. (1963) Texture profile method. *Journal of Food Science*, 28(4), 404–410.

Buisson, P.D. (1995) Developing new products for the consumer. In: D. Marshall (ed), *Food Choice and the Consumer*, Blackie Academic and Professional, London.

Cairncross, S.E. & Sjöstrom, L.B. (1950) Flavor profiles – a new approach to flavor problems. *Food Technology*, 4, 308–311.

Dairou, V. & Sieffermann, J.M. (2002) A comparison of 14 jams characterized by conventional profile and a quick original method, the flash profile. *Journal of Food Science*, 67, 826–834.

Department for Innovation, Universities and Skills (2007) Rigour, respect and responsibility: a universal ethical code for scientists. http://www.berr.gov.uk/files/file41318.pdf and http://www.berr.gov.uk/dius/science/science-and-society/public_engagement/code/page28030.html

Deschamps, J.P. & Nayak, P.R. (1996) *Product Juggernauts: How Companies Mobilise to Generate A Stream of Market Winners*. Harvard Business School Press, Boston, MA.

Dijksterhuis, G.B. (1997) *Multivariate Data Analysis in Sensory and Consumer Science*. Food & Nutrition Press Inc., Trumbull, Connecticut, USA.

Dijksterhuis, G. Flipsen, M. & Punter, P. (1994) PCA of TI-curves: three methods compared. *Food Quality and Preference*, 5, 121–127.

Ereaut, G., Imms, M. & Callingham, M. (2002) *Qualitative Market Research: Principle and Practice*. 7 Volumes. Sage Publications, London, UK.

Eriksson, L., Johansson, E., Kettaneh-Wold, N., Wikstrom, C. & Wold, S. (2000) *Design of Experiments: Principles and Practice*. Umetrics AB, Umea, Sweden.

EU (2001) European Parliament and Council Directive 2001/20/EC of 4 April 2001 on the approximation of the laws, regulations and administrative provisions of the Member States relating to the implementation of good clinical practice in the conduct of good clinical trials on medicinal products for human use. *Official Journal of the European Communities*, L121, 34–44, 1 May 2001.

EU (2005) European Parliament and Council Directive 2005/28/EC of 8 April 2005 laying down principles and detailed guidelines for good clinical practice as regards investigational medicinal products for human use, as well as the requirements for authorisation for the manufacturing or importation of such products. *Official Journal of the European Communities*, L91, 13–19, 9 April 2005.

Goldstein, E.B. (2006) *Sensation and Perception*. Wadsworth Publishing, Florence, KY.

Gordin, H.H. (1987). Intensity variation descriptive methodology: development and application of a new sensory evaluation technique. *Journal of Sensory Studies*, 2, 187–198.

International Chamber of Commerce and the European Society for Opinion and Marketing Research (2007) *ICC/ESOMAR International code of marketing and social research*. http://www.iccwbo.org

ISO 1987 ISO 8588: Sensory analysis of food. Part 5: A not A test.

ISO 1988 ISO 8587: Methods for sensory analysis of food. Part 6: Ranking.

ISO 1988 ISO 8589: Sensory analysis. General guidance for the design of test rooms.

ISO 1991 ISO 3972: Sensory analysis of food. Part 7: Investigating sensitivity of taste.

ISO 1992 ISO 5496: Sensory Analysis of food. Part 9: Initiation and training of assessors in the detection and recognition of odours.

ISO 1993 ISO 8586-1: Assessors for sensory analysis. Part 1: Guide to the selection, training and monitoring of selected assessors.

ISO 1994 ISO 8586-2: Assessors for sensory analysis. Part 2: Guide to the selection, training and monitoring of experts.

ISO 2000 ISO 9000: Quality Management Systems – Fundamentals and vocabulary.
ISO 2000 ISO 9001: Requirements.
ISO 2002 ISO 13301: Sensory analysis – Methodology – General guidance for measuring odour, flavour and taste detection thresholds by a three alternative forces choice (3-AFC) procedure.
ISO 2004 ISO 4120: Sensory analysis – Methodology – Triangle test.
ISO 2004 ISO 10399: Sensory analysis – Methodology – Duo-trio test.
ISO 2005 ISO 5495: Sensory analysis – Methodology – Paired comparison test.
ISO 2006 ISO 13300-1: Sensory analysis. General guidance for the staff of a sensory evaluation laboratory. Part 1: Staff responsibilities.
ISO 2006 ISO 13300-2: Sensory analysis. General guidance for the staff of a sensory evaluation laboratory. Part 2: Recruitment and training of panel leaders.
ISO 2006 ISO 20252: Market, opinion and social research. Vocabulary and service requirements.
Larson-Powers, N.M. & Pangborn, R.M. (1978) Descriptive analysis of the sensory properties of beverages and gelatine containing sucrose and synthetic sweeteners. *Journal of Food Science*, 43(11), 47–51.
Lawless, H.T. & Heymann, H. (1998) *Sensory Evaluation of Food: Principles and Practices*. Springer, New York, NY.
Lea, P., Naes, T. & Rødbotten, M. (1997) *Analysis of Variance for Sensory Data*. John Wiley & Sons Ltd., Chichester, England.
Liu, Y.H. & MacFie, H.J.H. (1990) Methods for averaging time–intensity curves. *Chemical Senses*, 15, 471–484.
MacFie, H.J.H. (2007) *Consumer Led Food Product Development*. Woodhead Publishing Limited, Cambridge, UK.
MacFie, H.J.H., Bratchell, N., Greenhoff, K. & Vallis, L.V. (1989) Designs to balance the effect of order of presentation and first-order carry-over effects in hall tests. *Journal of Sensory Studies*, 4(2), 129–148.
Meilgaard, M., Civille, C.V. & Carr, B.T. (2007) *Sensory Evaluation Techniques. Fourth Edition*. CRC, Boca Raton, FL.
Meullenet, J.F., Xiong, R. & Findlay, C. (2007) *Multivariate and Probabilistic Analyses of Sensory Science Problems*. Blackwell Publishing, Ames, IA, US.
Moskowitz, H.R. (1985) *New Directions for Product Testing and Sensory Analysis of Foods*. Food and Nutrition Press, Westport, CT.
Moskowitz, H.R., Beckley, J.H. & Resurreccion, A.V.A. (2006) *Sensory and Consumer Research in Food Product Design and Development*. Blackwell Publishing, Ames, IA.
Naes, T. & Risvik, E. (1996) *Multivariate Analysis of Data in Sensory Science (Data Handling in Science and Technology)*. Elsevier Science B.V., Amsterdam, The Netherlands.
Neilson, A.J., Ferguson, V.B., Kendall, D.A. (1988) Profile methods: flavor profile and profile attribute analysis. In: H.R. Moskowitz (ed), *Applied Sensory Analysis of Foods*. Vol. 1, Ch 2, pp. 21–41. CRC Press, Boca Raton, FL.
O'Mahony, M. (1986) *Sensory Evaluation of Food: Statistical Methods and Procedures*. Marcel Dekker, New York, NY.

O'Mahony, M. (1992) Understanding discrimination tests: A user friendly treatment of response bias, rating and ranking R-index tests and their relationship to signal detection. *Journal of Sensory Studies*, 9, 1–47.

Peryam, D.R. & Girardot, N. (1952) Advanced taste-test method. *Food Engineering*, 24(7), 58–61.

Pope, J.L. (1993) *Practical Marketing Research*. AMACOM, New York, pp. 295–308.

Poste, L.M., MacKie, D.A., Butler, G. & Larmond, E. (1991) *Laboratory Methods for Sensory Analysis of Food*. Canadian Publication Group Publishing Centre, Ottowa Canada, Publication 1864/E.

Rayner, J.C.W, Best, J., Brockhoff, P.B. & Rayner, G. (2006) *Nonparametrics for Sensory Science: A More Informative Approach*. Blackwell Publishing, Ames, IA, US.

Schutz, H.G. & Cardello, A.V. (2001) A labelled affective magnitude (LAM) scale for assessing food liking/disliking. *Journal of Sensory Studies*, 16, 117–159.

Stampanoni, C.R. (1994) The use of standardized flavor languages and quantitative flavor profiling technique for flavored dairy products. *Journal of Sensory Studies*, 9, 383–400.

Stone, H. & Sidel, J.L. (2004) *Sensory Evaluation Practices*, Academic Press, New York.

Stone, H., Sidel, J., Oliver, S., Woolsey, A. & Singleton, R.C. (1974) Sensory evaluation of quantitative descriptive analysis. *Food Technology*, 28(1), 24, 26, 28, 29, 32, 34.

Szczesniak, A.S., Brandt, M.A. & Friedman, H.H. (1963) Development of standard rating scales for mechanical parameters of texture and correlation between the objective and the sensory methods of texture evaluation. *Journal of Food Science*, 28, 397–403.

Szczesniak, A.S. (1963) Classification of textural characteristics. *Journal of Food Science*, 28, 385–389.

The Nuremberg Code (1949) In: *Trials of War Criminals Before the Nuremberg Military Tribunals Under Control Council Law No. 10*, Vol. 2, Nuremberg, October 1946–April 1949, pp. 181–182. Government Printing Office, Washington, DC.

Williams, A.A. & Langron, S.P. (1984) The use of free choice profiling for the examination of commercial ports. *Journal of the Science of Food and Agriculture*, 35, 558–568.

World Medical Association (2004) *Declaration of Helsinki*. http://www.wma.net

Index

2-alternative forced choice (2-AFC) test, 83–5
3-alternative forced choice (3-AFC) test, 85–7
α risk, 25, 70, 91, 92, 93, 153–4
β risk, 25, 91, 92, 93

'A' 'not A' test, 80–83, 176
absolute scale, 99
acceptance tests, 129–34
accuracy, 62, 102–3
action standards, 13–14
adaptation, 9
affective/consumer tests, 118–19
  general considerations, 119–20
  qualitative methods, 123
    focus groups, 123–6
  quantitative methods, 127
    acceptance tests, 129–34
    attribute diagnostics, 134–6
    preference tests, 127–9
  questionnaire design, 120
    coding, 122
    dos and don'ts, 123
    layout, 121
    pilot, 122
    question, type of, 121
    research objectives, 120
    type of, 121
    wording, 122
  training for, 61
  see also subjective tests
agreement scales, 135
air freshener, 1, 113
air handling, 45
allergens, 36
alpha risk, *see* α risk
American society for testing and materials (ASTM) standard, 67
analysis of variance (ANOVA), 16
  calculation, 153
  designs, 152–3
  interpretation, 153–5
  output for hardness, 168
  purpose, 151
  variation, sources of, 151–2
anchors, 100, 160

apple, 6, 14, 52, 70, 99
aroma, 5, 6, 53, 58, 59, 68, 110
  see also odour
assessment area, 43, 44, 46–7
  see also booths
assessment protocol, 36, 37, 98–9, 101
assessor–sample interaction plots
  for hardness, 169, 172
  for rate of melting, 169
assessors, 16, 54, 152
  for discrimination tests, 60
    affective tests, 61
    descriptive tests, 60–61
  dos and don'ts, 63
  good working practices for, 62
  internal vs. external assessors/panels, 55–6
  monitoring panel performance, 62–3
  motivation, 61
    feedback, 61
    group activities, 61–2
    personal contact, 61
    remuneration, 62
  protection, 31
  recruitment
    advertisement, 54–5
    direct recruitment, 55
    word of mouth/recommendation, 55
  screening and selection, 56
    naive assessors, 56–7
    trained panel, 57–60
  selection, 33, 97
  training, 60, 97
    affective tests, 61
    descriptive tests, 60–61
    discrimination tests, 60
attribute-specific tests, 83
  3-alternative forced choice, 85–7
  paired comparison (2-AFC), 83–5
  ranking test, 87–91
attribute(s)
  agreement on, 98
  diagnostic data, 134–6
  dumping, 8
  generation, 97–8
  intensity, 9, 135, 136

189

attribute(s) (*continued*)
  list, 164
  order of assessment, 111
audition, 6

balanced incomplete block (BIB) design, 17
balanced reference technique, 71
balanced test design, 15
baseline sample, *see* blank sample
beta risk, *see* β risk
beverages, 50, 51, 53, 58, 143, 145
binomial tests, 21
bipolar scale, 100
blank sample, 15
block designs, 16
blocks, 16
booths, 43, 46, 47, 166
  see also assessment area
bread, 11, 51
budget, 12

carrier, 51, 98
carry-over effects, 9
category-ratio scale, 160–61
  see also labelled magnitude scale (LMS)
category-specific scale, 99
central tendency, 23–4
central tendency error, 9
chemesthesis, 5
chewing gum, 98
chilli, 98
chi-squared test, 21, 77, 81
cleaning, 39
closed questions, 121
cluster analysis, 137
coding, 51–2, 122
coefficient of variation (CV), 102, 104
coffee, 1, 50, 113
colour blindness tests, 58
comparative/simultaneous designs, 17
complete block design, 16
completely randomised design (CRD), 16
computerised systems, for data capture, 64
confidence intervals, 24, 106, 107
conjoint analysis, 137
consensus profiling, 110
constant reference technique, 71
consumer tests, 2, 3, 12, 50, 51, 53, 66, 118, 119
consumers, 57
continuous time intensity, 116–17

contrast and convergence effects, 8
Control of Substances Hazardous to Health (COSHH) Regulations 2002, 38
control samples, 14–15
copy writing, 40
costs, 41, 64
crackers, 51, 98, 99
critical value, 25–6
critical values table
  for chi-square for degrees of freedom, 156
  for duo-trio test and paired comparison test, 149–50
  for Friedman test, 159
  for paired comparison and paired difference test, 157–8
  for triangle test, 147–8
cross-modal scaling, 162
cucumber, 99

data
  analysis, 19–29
  capture, 63
    computerised systems, 64
    paper, 63–4
    portable systems and internet, 64–5
    qualitative research, 65
  checks, 22
  distribution, 20
    nonnormal distribution, 20–21
    nonparametric tests, 21
    normal distribution, 20
    parametric tests, 21
  generation, 101
  handling, 22–3
    and documentation, 39–40
  interval, 19
  missing values, 22
  nominal, 19
  ordinal, 19
  outliers, 22
  quality, 102–5
  ratio, 19
  and record keeping, 39–40
  samples and population, 21–2
  skewed, 29
  storage, 140–41
  transformation, 22
    logarithmic transformation, 23
    normalising, 22
    standardising, 23
deodorant, 98
dependent samples, *see* related samples

descriptive analysis tests
  descriptive methodology, types of, 110–18
  determining objectives and future needs, 96
  dos and don'ts, 118
  key steps, 97
    agreement on attributes, 98
    assessment protocol, determining, 98–9
    attribute generation, 97–8
    data analysis and reporting, 102–9
    data generation, 101
    performance check, 101
    rating intensity, 99–101
    selection and training, of assessors, 97
  panel leader, role of, 96–7
  practical issues, in dealing with long-standing panels, 109
descriptive statistics, 21, 23
  central tendency, 23–4
  dispersion, 24–5
descriptive tests, 66
  training for, 60–61
design structure
  balanced test designs, 16
  common sensory designs, 16–17
  complete and incomplete block designs, 16
  panel size, 18–19
  randomisation, 15
  replication, 18
  sample presentation techniques, 17–18
difference from control profiling, 113
difference from control test, 73–6
discrete-point time intensity, 115–16
discrimination tests, 66–8
  attribute-specific tests, 83
    3-alternative forced choice, 85–7
    paired comparison (2-AFC), 83–5
    ranking test, 87–91
  dos and don'ts, 95
  forced choice vs. no difference, 67–8
  overall difference tests, 68
    'A' 'not A' test, 80–83
    difference from control test, 73–6
    duo-trio test, 71–3
    same–different test, 77–80
    triangle test, 68–70
  re-assessing samples, 67
  setting objectives, for test, 67
  similarity, 91

correct number of assessors, selecting, 92
power, of test, 91–2
triangle test to determine similarity, 92–5
true discriminators, proportion of, 92
test environment, 67
training for, 60
discussion area, 43
discussion guide, 124
discussion room, 48
dispersion, 24–5
distraction error, 7
distribution, of data
  nonnormal distributions, 20–21
  nonparametric tests, 21
  normal distribution, 20
  parametric tests, 21
documentation, 140
  and data handling, 39–40
  and data storage, 140–41
dummy sample, 15, 130
dumping, 98
duo-trio test, 71–3

equipments, 48, 49
Ethical Review Committee, 32
ethnography, 123
expectation error, 6
experimental/assessment procedure, safety of, 36–7
experimental design, 14
  design structure, 15–19
    balanced test designs, 16
    common sensory designs, 16–17
    complete and incomplete block designs, 16
    panel size, 18–19
    randomisation, 15
    replication, 18
    sample presentation techniques, 17–18
  dos and don'ts, 19
  treatment structure, 14
    control samples, 14–15
experimental error, 6
  cultural factors, 10
  physiological factors
    adaptation, 9
    physical condition, 10
    stimuli interactions, 9–10
  psychological factors
    attribute dumping, 8
    central tendency error, 9

experimental error (*continued*)
   contrast and convergence
      effects, 8
   distraction error, 7
   expectation error, 6
   habituation, 8
   halo effect, 7–8
   logical error, 7
   motivation error, 9
   order effect, 8
   proximity error, 7–8
   stimulus error, 7
   suggestion effect, 7
external vs. internal assessors/panels, 55–6

fatigue, 18, 74, 80
feedback, 61
flash profiling, 112
Flavor Profiling®, 110
focus group, 47, 119, 123–6
food products testing
   cleaning, 39
   hygiene, 38–9
   safety, 38
food texture, attributes of, 5
forced choice vs. no difference option, 67–8
free choice profiling, 112
frequency distribution, 20
Friedman test, 89, 136, 159
frozen desserts, 50
full written report, 138
functional areas
   assessment area (booths), 46–7
   discussion room, 48
   reception/waiting area(s), 45
   sample preparation area, 45–6
   sample serving area, 46
   specialised assessment area(s), 47
   storage areas, 48
   temporary assessment area(s), 47
   testing in the home (home use testing), 47–8

Gaussian distribution, *see* normal distribution, of data
generalised procrustes analysis (GPA), 112
good working and laboratory practices, 37
   documentation and data handling, 39–40
   dos and don'ts, 41
   food products testing, 38–9
   intellectual property, 40–41
   quality, 38
   safety, 38
   group activities, 61–2
   gustation, 4–5

habituation, 8
halo effect and proximity error, 7–8
hedonic rating, 129–34
home use testing (HUT), 47–8
hot food, 50
human senses, 4
   audition, 6
   gustation, 4–5
   multimodal perception, 6
   olfaction, 5
   touch, 5
   vision, 4
hygiene, 38–9
hypothesis
   alternative, 25
   null, 25, 26
   testing, 25–6

ICC/ESOMAR International Code of Marketing and Social Research, 31
IFST PFSG professional code of conduct for sensory professionals, 143–6
incomplete block design, 16
independent ethical committee review, 31
   assessors selection, 33
   compensation, 34
   health monitoring, 34
   informed consent, 32–3
inferential statistics, 21–2
   hypothesis testing, 25–6
informed consent, 32–3
Institute of Food Science and Technology (IFST), 31
intellectual property, 40–41
intensity
   range, 99–100
   rating, 99–101
   training, 100–101
intensity variation descriptive method, 113
interaction, 152
internal vs. external assessors/panels, 55–6
International organisation for standardisation (ISO), 67
interval data, 20
interview
   group, 123
   one-to-one, 61, 119, 121, 123

intra-oral perception, 6
ISO 20252:2006, 31

just about right (JAR) scales, 135

kinesthesis, 5

labelled magnitude scale (LMS), 160–61
Latin Square, 16
Latin Square designs, 142
legislation and professional codes of conduct, 31
lighting, 45
line scale, 160
logarithmic transformation, 23
logical error, 7

magnitude estimation, 160
magnitude matching, *see* cross-modal scaling
market assessment, 96
market research, 31
mean, 20
  arithmetic, 23, 136
  geometric, 23
median, 20, 23–4
melon, 52, 99
microbiological safety, in sensory testing, 35–6
missing values, 22
mode, 20, 24
modelling techniques, 137
modified quantitative descriptive analysis of chocolate texture, case study
  recruitment and training, 163–70
  sample evaluation, 170–73
monadic, 17
motivation error, 9
mouth-feel attributes, 5
multimodal perception, 6
multiple comparison tests (MCTs), 75, 153–4
  choice, 154–5

naive assessors
  consumers, 57
  untrained sensory panel, 56–7
no difference option vs. forced choice, 67–8
nominal data, 20
nonnormal distribution, of data, 20–21
nonparametric tests, 21, 27, 131, 136
normal distribution, of data, 20

normalising, 22
novel ingredients, 30, 35

objective tests, 1–2, 66
odour, 5, 38, 44, 45, 46, 47, 50, 58
  see also aroma
olfaction, 5
one-factor ANOVA, 152
open-ended questions, 121
oral presentation, 138
order effect, 8
ordinal data, 20
outlier, 22
overall difference tests, 68
  'A' 'not A' test, 80–83
  difference from control test, 73–6
  duo-trio test, 71–3
  same–different test, 77–80
  triangle test, 68–70

p-value, 25–6
paired comparison test (2-AFC), 83–5
paired preference test, 83
paired samples, *see* related samples
palate cleanser, 52
panel leader, role of, 96–7
panel performance, monitoring, 62–3
panel size, 18–19
parameters, definition of, 21
parametric tests, 20, 21, 27, 106, 160
partial least squares (PLS) regression, 137
patent, 40, 41
percentile ranges, 25
performance check, 101
perfumes, 51, 53
personal contact, 61
personal data, 140–41
population and samples, 21–2
portable systems and internet, 64–5
poster, 138
precision, 62, 104–5
preference mapping, 136–7
preference tests, 127–9
principal component analysis (PCA), 23, 108–9, 171
problem data, handling, 105–6
product assessment, 54, 98
product development, 96
product optimisation, 96
product-specific scale, *see* category-specific scale
product type, 11–12
professional conduct, in sensory testing
  dos and don'ts, 37
  experimental/assessment procedure, safety of, 36–7

194 Index

professional conduct, in sensory
    testing (*continued*)
  importance, 30
  independent ethical committee
    review, 31–4
  legislation and professional codes of
    conduct, 31
  protection of assessors, 31
  test samples, safety of, 34–6
Professional Food Sensory Group
  (PFSG), 31
Profile Attribute Analysis® (PAA®), 110
project completion
  data storage, 140–41
  documentation, 140
  dos and don'ts, 141
  reporting, 138–40
project planning
  action standards, setting, 13–14
  budget, 12
  data analysis, 19
    appropriate statistical test, choice
      of, 26–8
    data handling, 22–3
    descriptive statistics, 23–5
    distribution, 20–21
    dos and don'ts, 28–9
    inferential statistics, 25–6
    samples and population, 21–2
    types, 19–20
  experimental design, 14
    design structure, 15–19
    dos and don'ts, 19
    treatment structure, 14–15
  objectives, 11
  product type, 11–12
  test method selection, 12–13
  timings, 12
proto monadic, 17–18

qualitative methods, 123
  focus groups, 123–6
qualitative research, 65, 121
quality assurance (QA), 2, 74, 96
quality control (QC), 73, 74, 96
Quantitative Descriptive Analysis®,
  111, 163
quantitative flavour profiling (QFP), 113
quantitative methods, 127
  acceptance tests, 129–34
  attribute diagnostics, 134–6
  preference tests, 127–9
questionnaire design, 120
  coding, 122
  dos and don'ts, 123
  layout, 121

pilot, 122
question, type of, 121
research objectives, 120
type of, 121
wording, 122

R index, 78, 174
  calculation, 175–7
  discrimination tests with response
    bias, 174
  multiple sample assessments, 175
  result interpretation, 177
radar plots, 106
randomisation, 15
randomised complete block design
  (RCBD), 16–17
rank/rating, 162
ranking test, 87–91, 177
rating intensity, 99–101
ratio data, 20
raw data, 140
re-assessing samples, 67
reception/waiting area(s), 45
record keeping, 40
reference samples, 52
related samples, 27
relative scales, 99
reliability (validity), 63, 103–4
remuneration, 62
repeated measures design, 17, 152
replication, 18
reporting, 138–40
requirements, for sensory testing, 30
  assessors, 54
    dos and don'ts, 63
    good working practices for, 62
    internal vs. external assessors/
      panels, 55–6
    monitoring panel performance,
      62–3
    motivation, 61–2
    recruitment, 54–5
    screening and selection, 56–60
    training, 60–61
  data capture, 63
    computerised systems, 64
    paper, 63–4
    portable systems and internet, 64–5
    qualitative research, 65
  good working and laboratory
    practices, 37
    documentation and data
      handling, 39–40
    dos and don'ts, 41
    facilities testing food products,
      special considerations for, 38–9

intellectual property, 40–41
quality, 38
safety, 38
professional conduct, 30
  dos and don'ts, 37
  experimental/assessment
    procedure, safety of, 36–7
  importance, 30
  independent ethical committee
    review, 31–4
  legislation and professional
    codes, 31
  protection of assessors, 31
  safety of test samples, 34–6
resources needed, 41
  air handling, 45
  dos and don'ts, 48–9
  equipment, 48
  functional areas, 45–8
  general considerations, 43–4
  lighting, 45
  location, 44
  materials, 44
  sensory staff, 41–2
samples, 49
  dos and don'ts, 53–4
  preparation, 49–50
  presentation, 50–52
  reference samples, 52
  sample assessment procedure, 53
response surface modelling (RSM), 137
role, of sensory evaluation, 2–3

safety
  of experimental/assessment
    procedure, 36–7
  good working and laboratory
    practices, 38
  of test samples, 34
    allergens, 36
    microbiological safety, 35–6
    sample ingredients, 34–5
    toxicological safety, 35
same–different test, 77–80, 176
samples, 16, 49, 152
  assessment procedure, 53
  batches, 21
  dos and don'ts, 53–4
  microbial quality, 50
  and population, 21–2
  preparation, 45–6, 49
    equipment and utensils, 49
    materials, 49
    method, 49–50
  presentation, 17–18
    carrier, 51
  coding, 51–2
  number and order, 52
  palate cleanser, 52
  sample size and temperature, 50
  vessel, 50–51
  reference samples, 52
  serving area, 46
scale
  design, 99–101
  labeling, 100
scientific paper, 138
sensitivity, 9
sensometrics, 19
sensory and consumer testing, 2–3
sensory data, displaying, 106–9
sensory evaluation, meaning of, 1
sensory lexicon, 98, 164, 165
sensory perception
  human senses, 4
    audition, 6
    gustation, 4–5
    multimodal perception, 6
    olfaction, 5
    touch, 5
    vision, 4
  sensory measurements, factors
    affecting, 6
    cultural factors, 10
    physiological factors, 9–10
    psychological factors, 6–9
sensory profiles, 106–7
sensory space, 99
sensory staff, 41–2
sensory test methods, 3, 151
  affective/consumer tests, 118–19
    general considerations, 119–20
    qualitative methods, 123–6
    quantitative methods, 127–36
    questionnaire design, 120–23
  consumer, sensory and product
    data, linking, 136
    conjoint analysis, 137
    modelling techniques, 137
    preference mapping, 136–7
  descriptive analysis tests, 96–7
  descriptive methodology, types
    of, 110–18
    dos and don'ts, 118
    key steps, 97–109
    practical issues, in dealing with
      long-standing panels, 109
  discrimination tests, 66–8
    attribute-specific tests, 83–91
    dos and don'ts, 95
    overall difference tests, 68–83
    similarity, 91–5

sensory test methods (*continued*)
  objective tests, 66
  subjective tests, 66
sensory traces, 106–7
sequential monadic, 17
short written report, 138
similarity, 91
  correct number of assessors, selecting, 92
  power, of test, 91–2
  triangle test to determine similarity, 92–5
  true discriminators, proportion of, 92
skewed data, 29
skin cream, 98, 124, 126
somesthesis, 5
specialised assessment area(s), 47
spectrum™ method, 111–12
spider plots, 106
standard deviation, 24
standard error, 24–5
standard operating procedures (SOPs), 38
standardisation, 23
star charts, 106
statistical test, choice of, 26–8
statistics
  definition, 21
  descriptive, 21, 23
    central tendency, 23–4
    dispersion, 24–5
  inferential, 21–2
    hypothesis testing, 25–6
stimuli interactions, 9–10
stimulus error, 7
storage areas, 48
students t-test, 21
study-specific scale, 99–100
subjective tests, 2, 66
suggestion effect, 7
sureness rating, 78, 80, 82
swallowing, 53

tea, 52
temperature, 50
temporal dominance of sensations (TDS), 117–18
temporary assessment area(s), 47
test environment, 67
test locations
  advantages, 120
  disadvantages, 120
test method selection, 12–13
test samples, safety of, 34
  allergens, 36
  microbiological safety, 35–6
  sample ingredients, 34–5
  toxicological safety, 35
testing facilities
  air handling, 45
  functional areas, 45–8
  general considerations, 43–4
  lighting, 45
  location, 44
  materials, 44
texture, 110
Texture Profiling®, 110–11
threshold tests, 58
time intensity methods, 113–15
timings, 12
toothpaste, 53
touch, 5
toxicological safety, in sensory testing, 35
trademarking, 40
trained panel
  health, 57
  personality, 57
  results interpretation, 59–60
  sensory acuity, 57–8
  stimuli
    detection, 58
    discriminating between, 58
    recognising and describing, 58–9
transformation, 22
  logarithmic transformation, 23
  normalising, 22
  standardising, 23
treatment structure, 14–15
triangle test, 68–70
  to determine similarity, 92–5
true discriminators, proportion of, 92
two-factor ANOVA, 152
  with interaction, 153
type I error, 25
type II error, 25

universal scale, 99
untrained sensory panel, 56–7

variation, 151–2
vision, 4
visual analogue scale, *see* line scale

water, 50, 99
Williams Latin Square designs, 142
wine, 5, 51, 116

yoghurt, 96, 99